AF389459

Condensed-Matter-Principia Based Information & Statistical Measures

Condensed-Matter-Principia Based Information & Statistical Measures

From Classical to Quantum

Editors

Adam Gadomski
Sylwia Zielińska-Raczyńska

MDPI • Basel • Beijing • Wuhan • Barcelona • Belgrade • Manchester • Tokyo • Cluj • Tianjin

Editors
Adam Gadomski
UTP University of Science and
Technology
Poland

Sylwia Zielińska-Raczyńska
UTP University of Science and
Technology
Poland

Editorial Office
MDPI
St. Alban-Anlage 66
4052 Basel, Switzerland

This is a reprint of articles from the Special Issue published online in the open access journal *Entropy* (ISSN 1099-4300) (available at: https://www.mdpi.com/journal/entropy/special_issues/cm_stat).

For citation purposes, cite each article independently as indicated on the article page online and as indicated below:

LastName, A.A.; LastName, B.B.; LastName, C.C. Article Title. *Journal Name* **Year**, *Article Number*, Page Range.

ISBN 978-3-03936-746-7 (Hbk)
ISBN 978-3-03936-747-4 (PDF)

Contents

About the Editors

Adam Gadomski, a full professor of physics at the University of Science and Technology, Bydgoszcz, Poland, leads a group of academics focused on the modeling of physicochemical processes. He specializes in statistical soft-condensed matter physics, and works in computational physics and physical computation problems, with an emphasis placed on biomaterials with a structure–property relationship.

Sylwia Zielińska-Raczyńska, a professor of physics at the University of Science and Technology, Bydgoszcz, Poland, is the leader of a group of academics focused on quantum optics and quantum information. She specializes in quantum optics and the nanophysics of (Rydberg) excitons, nanostructures and metamaterials, with a special emphasis on quantum information issues.

Preface to "Condensed-Matter-Principia Based Information & Statistical Measures"

The recent development of research in the fields related to quantum information has become one of the most intriguing investigations in contemporary physics. Many novel concepts are discussed nowadays in the literature, and the broad range of new models of quantum optics and solid-state physics have recently been considered in the context of quantum information theory. A lot of effort has been devoted to investigating new ideas connected to various aspects of quantum correlations, such as entanglement and quantum steering, but also to more practical proposals of the systems, which could be applied in quantum technology. Hereafter, we will juxtapose the main message(s) from ten papers published in the underlying Special Issue (SI).

The book content is primarily based on the efforts gathered by the colloquium "What is INFORMATION—how to get, process, code and read it, in the elegant world of mathematical, statistical and quantum physics", which took place on 24–25 October, 2019, at the University of Science and Technology (UTP) in Bydgoszcz, Poland. It was devoted, in part, to recognizing the outstanding contribution to statistical thermodynamics and condensed matter physics by Professor Gerard Czajkowski, former Director of Institute of Mathematics and Physics and Vice Rector for Research. Gerard Czajkowski obtained his master's degree in 1967, his Ph.D. in 1971, and his habilitation in theoretical physics in 1976, all from Nicolaus Copernicus University in Toruń, Poland. He became a full professor of physical sciences in 1987. All his thesis' topics of interests were devoted to the statistical aspects of thermodynamics paving the way for quantum information. Over the fifty years of his academic career, his research interests have evolved from information–theoretical methods in the stochastic theory of open systems, to the quantum optics of semiconductors unlocking a plethora of dynamical effects which might be useful for the practical realization of quantum information processing in solids.

This book has accumulated substantial and versatile evidence on the widely distributed phenomenon of information transmission, and how to measure it from a number of fairly complex systems, ranging from classical to quantum (albeit discussed purposely in a reversed order), that were thoroughly studied by the contributions collected in the book. The studies by Professor Gerard Czajkowski and coworkers enabled, to a great extent, the accomplishment of the goal, by extracting many useful information measures from the systems covered by the book's contents.

Adam Gadomski, Sylwia Zielińska-Raczyńska

Editors

Editorial

Information and Statistical Measures in Classical vs. Quantum Condensed-Matter and Related Systems

Adam Gadomski [1,*] **and Sylwia Zielińska-Raczyńska** [2]

[1] Institute of Mathematics and Physics (Group of Modeling of Physicochemical Processes), Faculty of Chemical Technology and Engineering, UTP University of Science and Technology, 85-796 Bydgoszcz, Poland

[2] Institute of Mathematics and Physics (Group of Quantum Optics and Information), Faculty of Chemical Technology and Engineering, UTP University of Science and Technology, 85-796 Bydgoszcz, Poland; sziel@utp.edu.pl

[*] Correspondence: agad@utp.edu.pl; Tel.: +48-523-408-697

Received: 6 June 2020; Accepted: 9 June 2020; Published: 10 June 2020

Abstract: The presented editorial summarizes in brief the efforts of ten (10) papers collected by the Special Issue (SI) "Condensed-Matter-Principia Based Information & Statistical Measures: From Classical to Quantum". The SI called for papers dealing with condensed-matter systems, or their interdisciplinary analogs, for which well-defined classical statistical vs. quantum information measures can be inferred while based on the entropy concept. The SI has mainly been rested upon objectives addressed by an international colloquium held in October 2019, at the University of Science and Technology (UTP) Bydgoszcz, Poland (see http://zmpf.imif.utp.edu.pl/rci-jcs/rci-jcs-4/), with an emphasis placed on the achievements of Professor Gerard Czajkowski (PGC). PGC commenced his research activity with diffusion-reaction (open) systems under the supervision of Roman S. Ingarden (Toruń), a father of Polish synergetics, and original thermodynamic approaches to self-organization. The active cooperation of PGC mainly with German physicists (Friedrich Schloegl, Aachen; Werner Ebeling, Berlin) ought to be underlined. Then, the development of Czajkowski's research is worth underscoring, moving from statistical thermodynamics to solid state theory, pursued in terms of nonlinear solid-state optics (Franco Bassani, Pisa), and culminating very recently with large quasiparticles, termed Rydberg excitons, and their coherent interactions with light.

Keywords: information; entropy; classical vs. quantum system; condensed matter; soft matter; complex systems

---

The colloquium "What is INFORMATION—How to get, process, code and read it, in the elegant world of mathematical, statistical and quantum physics" which took place on 24–25 October 2019 at the University of Science and Technology (UTP) in Bydgoszcz, Poland, was devoted in part to recognizing the outstanding contribution to statistical thermodynamics and condensed matter physics by Professor Gerard Czajkowski, former director of the Institute of Mathematics and Physics and vice rector for research. Gerard Czajkowski obtained a master's degree in 1967, Ph.D. in 1971, and habilitation in theoretical physics in 1976, all from Nicolaus Copernicus University in Toruń, Poland. He became a full professor of physics in 1987. All of his thesis topics of interest were devoted to the statistical aspects of thermodynamics, paving the way for quantum information. During fifty years of his academic career, his research interests have evolved from information-theoretical methods in the stochastic theory of open systems to the quantum optics of semiconductors, unlocking a plethora of dynamical effects which might be useful for the practical realization of quantum information processing in solids.

Recent research developments in the fields related to quantum information have become some of the most intriguing investigations in contemporary physics. Many novel concepts nowadays are discussed in the literature, and a broad range of new models of quantum optics and solid-state physics have recently been considered in the context of quantum information theory. A lot of effort has been devoted to investigating new ideas connected to various aspects of quantum correlations, such as entanglement and quantum steering, but also to more practical proposals of the systems, which could be applied in quantum technology. Here, let us juxtapose the main message(s) from ten papers published in the underlying Special Issue (SI).

The paper "Binary Communication with Gazeau–Klauder Coherent States" [1] investigates the advantages and disadvantages of using Gazeau–Klauder coherent states for optical communication. It is proved that using an alphabet consisting of coherent Gazeau–Klauder states related to a Kerr-type nonlinear oscillator instead of standard Perelomov coherent states results in a lowering of the Helstrom bound for error probability in binary communication. The authors also discuss the trace distance between Gazeau–Klauder coherent states and a standard coherent state as a quantifier of distinguishability of alphabets.

Photons are excellent carriers of information, but they are essentially noninteracting with each other in the absence of a dispersive medium. In order to overcome this difficulty, photons should be localized at the characteristic length scale in the medium which enables mapping photons into medium excitations, preserving their phase relations. Electromagnetically induced transparency (EIT) is a phenomenon which enables the transference of coherence and quantum states from light to collective medium spin excitations and has turned out to be very attractive for applications in quantum information. Solid bulk media are systems well worth considering for storing and processing quantum information because they have a number of advantages over atomic gases. Recently, Rydberg excitons have attracted a great deal of attention due to their exciting features: the distinct combination of their long radiative lifetimes, sensitivity to external fields, and strong dipolar interactions could be exploited to realize quantum interfaces for quantum information processing.

In the paper entitled "Electromagnetically Induced Transparency in Media with Rydberg Excitons 1: Slow Light" [2], it is shown that electromagnetically induced transparency (EIT) can be realized in media with Rydberg excitons. With realistic, reliable parameters which show good agreement with optical and electro-optical experiments, as well as the proper choice of Rydberg exciton states in cuprous oxide crystal, the author indicates how EIT can be achieved. The calculations show that, due to a large group index, one can expect the slowing down of a light pulse by a factor of about 10^4 in this medium.

The paper "Electromagnetically Induced Transparency in Media with Rydberg Excitons 2: Cross-Kerr Modulation" [3] discusses the issue of nonlinear interaction of photons in the linear regime of EIT. By mapping photons into the sample of cuprous oxide with Rydberg excitons, it is possible to obtain a significant optical phase shift due to third-order cross-Kerr nonlinearities realized in a transparent medium. The optimum conditions for observation of the phase shift over π in Rydberg excitons media are examined. A discussion of the application of the cross-phase modulations in the field of all-optical quantum information processing in solid-state systems is presented.

The proposal of matching fractal plasmons, a quantum or quasiparticle associated with a local collective oscillation of charge density, with information processing is discussed in "Fractal Plasmons on Cantor Set Thin Film" [4]. The propagation of surface plasmon–polaritons is investigated in a metallic, fractal-like structure based on the Cantor set. This work uncovers a novel opportunity to elicit information about the surface structure from the reflection spectrum of surface plasmon–polaritons and thereby establishes a link between optical properties and different fractal geometries.

Another paper referring to plasmonics, "Interaction and Entanglement of a Pair of Quantum Emitters near a Nanoparticle: Analysis beyond Electric-Dipole Approximation" [5], considers the problem of interactions between quantum particles positioned near plasmonic nanostructures which may be substantially stronger than in free space, leading to fast energy exchange between the particles and corresponding fast entanglement generation. The authors study the influence on these quantities of interaction channels beyond the electric dipole. It may not only be significant in terms of numbers, but may also affect the spatial symmetry of the optical response as a function of the position of particles with respect to the plasmonic nanostructure.

The first of the classical statistical papers [6] of the underlying SI is about the profound misuses, misunderstanding, misinterpretations, and misapplications of entropy, the Second Law of Thermodynamics and Information Theory. It points, in the author's opinion [6], to the "Greatest Blunder Ever in the History of Science". It is not about a single blunder admitted by a single person (e.g., Albert Einstein's assertions on the cosmological constant), but more a blunder of immense proportions, the claws of which have spread over all branches of science; from thermodynamics, cosmology, and biology to sociology and much more.

The second from the series of classical statistical papers [7] embarks on a versatile analysis of the conformation of albumin in the temperature range of 300–312 K, i.e., in the physiological range. By employing molecular dynamics simulations, values of the backbone and dihedral angles for this molecule are calculated. An analysis is performed on the global dynamic properties of albumin treated as a chain. In this temperature range, the parameters of the molecule and the conformational entropy derived from two angles that reflect global dynamics in the conformational space are obtained. A rationale based on scaling theory for the subdiffusion Flory–De Gennes type exponent of 0.4 is unfolded in conjunction with detecting the most appreciable fluctuations of the corresponding statistical test parameter. These fluctuations are shown to coincide adequately with entropy fluctuations, namely, the oscillations around the thermodynamic equilibrium. By applying the Kullback–Leibler theory, differences between the distribution of the root-mean-square displacement for each temperature and time window are addressed.

The third from the series of classical statistical papers [8] regards the problem of information processing toward living organisms while based on chemical reactions. It is worth noting that the human achievements in constructing chemical information processing devices demonstrate that it is difficult to design such devices using the bottom-up approach. Therefore, an alternative top-down design of a network of chemical oscillators that performs a selected computing task is worth pursuing. As an example, a simple network of interacting chemical oscillators that operates as a comparator of two real numbers is analyzed. The information on which of the two numbers is larger is coded in the number of excitations detected on oscillators forming the network. The parameters of the network are optimized to perform this function with maximum accuracy. A discussion is carried out on how information theory methods can be applied to obtain the optimum computing structure.

The next from the series of classical statistical papers [9] studies the flow of a substance in a channel network which consists of nodes of the network and edges which connect these nodes and form paths for motion of the substance. The channel can have an arbitrary number of arms, and each arm can contain an arbitrary number of nodes. The flow of the substance is modeled by a system of ordinary differential equations. First, a model for a channel with arms each involving an infinite number of nodes is discussed. For the stationary regime of motion of a substance in such a channel, the probability distributions connected to the distribution of the substance in any of channel's arms and in the entire channel are derived. The obtained distributions can be connected to the Waring distribution. Next, a model for the flow of a substance in a channel with arms each containing a finite number of nodes is analyzed. The probability distributions connected to the distribution of the substance in the nodes of the channel for the stationary regime of flow of the substance are obtained. These distributions are new, and based on them the corresponding information measure and Shannon information measure for the studied kind of flow of a substance are derived.

The last work from the series of classical statistical papers [10] collected by the SI addresses in a biomimetic way the subject of drones and drone swarms, equipped with sophisticated algorithms that help them achieve mission objectives. Such algorithms vary in their quality such that their comparison needs a metric that would allow for their correct assessment. The novelty of this study relies on analyzing, defining, and applying to swarms the construct of cross-entropy, known from thermodynamics and information theory. It can be used as a synthetic measure of the robustness of algorithms that can control swarms in the case of unexpected obstacles and unforeseen problems. This work applies in terms of necessary formalizations (algorithms) and, by addressing a few examples, to collision avoidance issues when prompted to react to material obstacles.

To sum up, this SI accumulates substantial and versatile evidence on the widely distributed phenomenon of information transmission and how to measure it from a number of fairly complex entropic systems, ranging from classical to quantum (albeit discussed purposely in a reversed order), that have been thoroughly studied by the contributors of the presented SI [1–10]. The studies by PGC and coworkers have, to a great extent, enabled the extraction of many useful information measures from the systems covered by the SI.

Funding: This research received a funding from UTP University of Science and Technology, Bydgoszcz, Poland, grant BN-10/19 (AG) and also from the National Science Centre, Poland (project OPUS 2017/25/B/ST3/00817; SZ-R).

Acknowledgments: We are very much indebted to Lidia Smentek (Toruń/Nashville) and Werner Ebeling (Berlin) for providing us with very interesting historical materials deposited at http://zmpf.imif.utp.edu.pl/rci-jcs/rci-jcs-4/. The Colloquium *per se* was supported by the Marshal of Kujawsko-Pomorskie Voivodeship in Toruń, Poland.

Conflicts of Interest: The authors declare no conflict of interest.

References

1. Dajka, J.; Łuczka, J. Binary Communication with Gazeau–Klauder Coherent States. *Entropy* **2020**, *22*, 201. [CrossRef]
2. Ziemkiewicz, D. Electromagnetically Induced Transparency in Media with Rydberg Excitons 1: Slow Light. *Entropy* **2020**, *22*, 177. [CrossRef]
3. Ziemkiewicz, D.; Zielińska - Raczyńska, S. Electromagnetically Induced Transparency in Media with Rydberg Excitons 2: Cross-Kerr Modulation. *Entropy* **2020**, *22*, 160. [CrossRef]
4. Ziemkiewicz, D.; Karpiński, K.; Zielińska-Raczyńska, S. Fractal Plasmons on Cantor Set Thin Film. *Entropy* **2019**, *21*, 1176. [CrossRef]
5. Kosik, M.; Słowik, K. Interaction and Entanglement of a Pair of Quantum Emitters near a Nanoparticle: Analysis beyond Electric-Dipole Approximation. *Entropy* **2020**, *22*, 135. [CrossRef]
6. Ben-Naim, A. Entropy and Information Theory: Uses and Misuses. *Entropy* **2019**, *21*, 1170. [CrossRef]
7. Weber, P.; Bełdowski, P.; Domino, K.; Ledziński, D.; Gadomski, A. Changes of Conformation in Albumin with Temperature by Molecular Dynamics Simulations. *Entropy* **2020**, *22*, 405. [CrossRef]
8. Gorecki, J. Applications of Information Theory Methods for Evolutionary Optimization of Chemical Computers. *Entropy* **2020**, *22*, 313. [CrossRef]
9. Borisov, R.; Dimitrova, Z.I.; Vitanov, N.K. Statistical Characteristics of Stationary Flow of Substance in a Network Channel Containing Arbitrary Number of Arms. *Entropy* **2020**, *22*, 553. [CrossRef]
10. Cofta, P.; Ledziński, D.; Śmigiel, S.; Gackowska, M. Cross-Entropy as a Metric for the Robustness of Drone Swarms. *Entropy* **2020**, *22*, 597. [CrossRef]

Article

Binary Communication with Gazeau–Klauder Coherent States

Jerzy Dajka [1,2,3,*] **and Jerzy Łuczka** [1,3,4]

[1] Institute of Physics, University of Silesia in Katowice, 40-007 Katowice, Poland; jerzy.luczka@us.edu.pl
[2] Institute of Computer Science, University of Silesia in Katowice, 40-007 Katowice, Poland
[3] Silesian Center for Education and Interdisciplinary Research, University of Silesia in Katowice, 40-007 Chorzów, Poland
[4] Institute of Mathematics, University of Silesia in Katowice, 40-007 Katowice, Poland
[*] Correspondence: jerzy.dajka@us.edu.pl

Received: 19 December 2019; Accepted: 6 February 2020; Published: 10 February 2020

Abstract: We investigate advantages and disadvantages of using Gazeau–Klauder coherent states for optical communication. In this short paper we show that using an alphabet consisting of coherent Gazeau–Klauder states related to a Kerr-type nonlinear oscillator instead of standard Perelomov coherent states results in lowering of the Helstrom bound for error probability in binary communication. We also discuss trace distance between Gazeau–Klauder coherent states and a standard coherent state as a quantifier of distinguishability of alphabets.

Keywords: Gazeau–Klauder coherent states; Helstrom bound

---

1. Introduction

Quantum optical implementations of quantum information processing, including communication and computation, seems to be one of the most promising kinds today [1]. It is related to the maturity of both theoretical and experimental techniques developed in the last hundred years. It was quite early when the Quantum Community recognized the usefulness of the 'most classical' among quantum states—the coherent states—in quantum information processing [1–4]. Even recently coherent states with a non-random phase, despite certain limitations [2], have found their application in the very hot branch of quantum communication related to quantum key distribution [5]. The idea is simply to utilize as an alphabet a pair of coherent states [1]

$$\rho_0 = |0\rangle\langle 0|, \quad \rho_1 = |z\rangle\langle z|, \tag{1}$$

where $|z\rangle = D(z)|0\rangle$ is a coherent state related to the vacuum state $|0\rangle$ via the displacement operator $D(z) = \exp\left(-za^\dagger - \bar{z}a\right)$ representing the Heisenberg–Weyl algebra $[a, a^\dagger] = 1$ [6]. Let us notice that the apparent simplicity of that proposal is at a price of non-orthogonality of the 'letters', i.e., $\mathrm{tr}(\rho_0\rho_1) \neq 0$, resulting in their limiting distinguishability. Since coherent states do not require nonlinear media for their generation it seems advantageous [3] to use them in comparison to, e.g., earlier proposals utilizing squeezed states [7] demanding 'hard' nonlinearity. However, recent progress in experimental techniques may reverse this trend at least in the cases when going beyond standard coherent states becomes advantageous. Using the Schrödinger cat states as candidates for orthogonal letters of alphabet states serves as an example [1].

The aim of this work is to present an example of a candidate for an alphabet consisting of Gazeau–Klauder coherent states [8]. We analyze binary communication with Gazeau–Klauder states related to an oscillator equipped with a polynomial nonlinearity typical for the Kerr media. The Gazeau–Klauder coherent states have been studied for a variety of quantum systems: A one-mode system with sinusoidal potential [9], systems characterized by the Pöschl-Teller [10] and Morse potentials [11], for nonlinear Kerr-type oscillators [12], quantum particles confined by a double–well potential [13] and pseudoharmonic oscillators [14]. They can also serve as a basis of a very natural generalization of the cat states [15,16]. Our aim is to expand a list of potential applications of Gazeau–Klauder construction to a class of communication problems utilizing the non-orthogonal binary alphabet formed by Gazeau–Klauder coherent states. We show that such a choice can result in lowering of the Helstrom bound for a receiver error that may balance an obvious disadvantage of using nonlinear systems leading to non-trivial Gazeau–Klauder coherent states. The paper is organized as follows: After providing a short review of the Gazeau–Klauder construction of coherent states for a quantum bosonic system with a Kerr-type polynomial nonlinearity we calculate, as a quantifier of distinguishability of states, trace distance between Gazeau–Klauder states for a nonlinear system and corresponding standard states in a linear limit. Further, we propose a binary communication scheme utilizing an alphabet consisting of two Gazeau–Klauder coherent states as an alternative for well established schemes utilizing standard (Perelomov) coherent states. For such a scheme we calculate the Helstrom bound minimizing (over all possible positive-operator-valued measurements (POVM)) the error in the receiver. In the last two sections we discuss and conclude our work.

2. Results

Standard coherent states [6] are most natural for harmonic potential systems exhibiting the Heisenberg–Weyl symmetry which is a first step toward generalized coherent states in the Gilmore-Perelomov sense [6] exhibiting different symmetries. However, in an absence of almost any symmetry it is still possible to construct a class of states equipped with most of the desired properties of coherent states: The Gazeau–Klauder coherent states [8] solely associated with Hamiltonians of systems under consideration.

For the sake of completeness, let us recall the construction proposed in reference [8]. Let H be a Hamiltonian of the system with purely discrete non-degenerate (either finite or infinite) spectrum. The first step in constructing the Gazeau–Klauder states is to solve the eigenvalue problem:

$$H|n\rangle = E_n|n\rangle \equiv \hbar\omega\varepsilon_n|n\rangle, \quad n = 0, 1, 2, \ldots \tag{2}$$

The Gazeau–Klauder coherent states $|J, \gamma\rangle$ are two-parameter states with real-valued $J \geq 0$ and $\gamma \in (-\infty, \infty)$ defined by the relation [8]

$$|J, \gamma\rangle = \frac{1}{C(J)} \sum_{n=0}^{\infty} \frac{J^{n/2} \exp(-i\gamma e_n)}{\sqrt{\rho_n}} |n\rangle, \tag{3}$$

where

$$e_n = \varepsilon_n - \varepsilon_0 = \frac{E_n - E_0}{\hbar\omega}, \quad \rho_n = \Pi_{j=1}^{n} e_j, \quad C^2(J) = \sum_{n=0}^{\infty} \frac{J^n}{\rho_n} \tag{4}$$

and $\rho_0 = 1$. The other parameters in Equation (3) can be equipped with a clear physical meaning [8]: (i) $\langle J, \gamma|H|J, \gamma\rangle \sim J$ corresponds to a mean energy of the system, and (ii) its phase γ is related to a temporal stability via $\exp(-iHt)|J, \gamma\rangle = |J, \gamma + \omega t\rangle$. As the maximal value of J is bounded from above by the radius of convergence of the series $C(J)$, a choice of J leading to a well defined quantum state remains limited.

The most elementary generalization of the standard coherent states [6] leading to the Gazeau–Klauder coherent states of non-trivial properties is for a bosonic oscillator with a polynomial nonlinearity. Let us consider a nonlinear oscillator of the Kerr type studied in the context of Gazeau–Klauder states in reference [12]. It is described by the bosonic Hamiltonian

$$H = \hbar\omega a^\dagger a + \hbar\chi a^{\dagger 2} a^2 \equiv \hbar\omega \hat{N} + \hbar\chi(\hat{N}^2 - \hat{N}), \tag{5}$$

where $a^\dagger$ and a are the creation and annihilation boson operators, $\hat{N} = a^\dagger a$ is a number operator and χ is related to the nonlinear susceptibility of the Kerr medium [12], i.e., a medium with a refraction index depending on the field intensity [17,18].

From Equation (5) it follows that the energy eigenvalues are given by

$$e_n = \varepsilon_n = n - \mu n + \mu n^2, \quad n = 0, 1, 2 \ldots \tag{6}$$

where $\mu = \chi / \omega$ (typically not exceeding a range of unity $\mu \sim 1$) is the susceptibility rescaled with respect to the bare oscillator energy and hence one can explicitly construct Gazeau–Klauder states with

$$\rho_n = \Gamma(n+1)\,\mu^n \Gamma\left(\frac{\mu n + 1}{\mu}\right) / \Gamma\left(\mu^{-1}\right). \tag{7}$$

Let us notice that $\rho_n = \Gamma(n+1) = n!$ for the harmonic oscillator, i.e., for $\mu = 0$. In this case the Gazeau–Klauder coherent states reduce to the standard coherent states for $z = \sqrt{J}$. Moreover, since eigenstates of the Kerr Hamiltonian (5) coincide with the eigenstates of the standard harmonic oscillator one can consider the Gazeau–Klauder states studied in this paper as a very first 'extension' of the standard coherent states construction of Perelomov [6]. In particular, for $\gamma = 0$, one gets

$$|J,0\rangle = \frac{1}{C(J)} \sum_{n=0}^{\infty} \frac{J^{n/2}}{\sqrt{\rho_n}} |n\rangle, \tag{8}$$

with

$$C^2(J) = I_{\frac{1-\mu}{\mu}}\left(2\sqrt{\frac{J}{\mu}}\right) \Gamma\left(\mu^{-1}\right) \left(\frac{J}{\mu}\right)^{-\frac{1-\mu}{2\mu}} \tag{9}$$

expressed in terms of the modified Bessel function $I_\alpha(z)$. Further, as a potential alternative for the traditional choice given by Equation (1), here we consider

$$\rho_0 = |0\rangle\langle 0|, \quad \rho_1 = |J,0\rangle\langle J,0| \tag{10}$$

as a candidate for the alphabet in a binary communication and compare it with Alphabet (1) with real $z \in \mathbf{R}$, cf. Reference [3].

For a binary communication utilizing two states in either Equation (1) or Equation (10) as codewords, a receiver is faced with a decision of distinguishing which among two states has already been transmitted. The most natural quantifier of the distinguishability between two various states ρ and σ is the trace distance [19]:

$$D = \frac{1}{2}\mathrm{Tr}\left[\sqrt{(\rho - \sigma)^2}\right]. \tag{11}$$

The trace distance between the Gazeau–Klauder coherent state and the corresponding standard coherent state satisfying $z = \sqrt{J}$ can be calculated and the result is

$$D = \sqrt{1 - |F|^2}, \tag{12}$$

where

$$F = \frac{e^{-J/2}}{C(J)} \sum_{n=0}^{\infty} J^n \sqrt{\frac{\Gamma\left(\mu^{-1}\right)}{\mu^n \Gamma^2(n+1)\,\Gamma\left(\frac{\mu n + 1}{\mu}\right)}} \tag{13}$$

stands for fidelity [19]. Clearly, with increasing μ which quantifies the role played by polynomial nonlinearity of the Kerr medium in Equation (5) the trace distance becomes larger as presented in Figure 1. Nevertheless, for small values of J, the Gazeau–Kluder and Perelomov coherent states are hardly distinguishable.

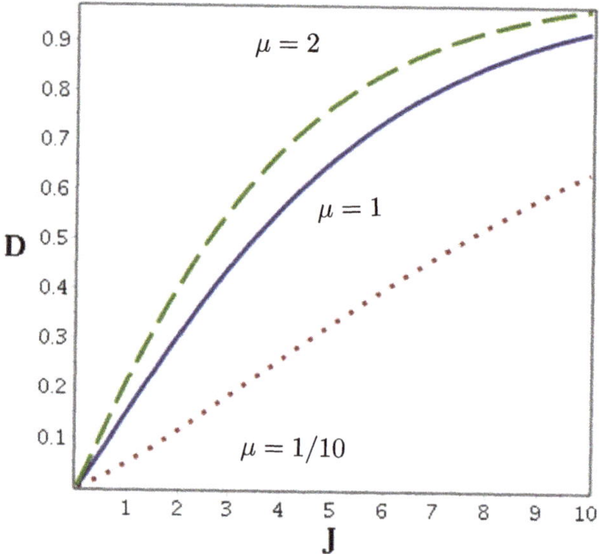

Figure 1. Trace distance between the Gazeau–Klauder coherent states $|J,0\rangle$ given by Equation (3) and the Perelomov coherent states $|z = \sqrt{J}\rangle$ depicted for selected values of the rescaled susceptibility μ.

Non-orthogonality of states ρ_0 and ρ_1 utilized as codewords in (binary) quantum communication becomes a natural source of error due to limited distinguishability of codewords. If one applies (resolving unity) POVM (positive-operator-valued measures) [19]

$$Id \;=\; \Pi_0 + \Pi_1 \tag{14}$$

for a measurement of non-orthogonal states one arrives to two hypotheses H_0 and H_1 which need to be tested. According to H_0, the transmitted state is ρ_0 and according to H_1, the transmitted state is ρ_1. There is also the natural and unavoidable possibility of erroneous detection and choosing H_0 (H_1) if ρ_1 (respectively, ρ_0) arrives at a receiver. Such an opportunity can be formalized by quantities

$$p(H_0|\rho_1) \;=\; \mathrm{tr}[\Pi_0\rho_1], \;\; p(H_1|\rho_0) = \mathrm{tr}[\Pi_1\rho_0]. \tag{15}$$

The receiver error probability becomes then

$$p[\Pi_0, \Pi_1] \;=\; p_0(\rho_0)p(H_1|\rho_0) + p_0(\rho_1)p(H_0|\rho_1) \tag{16}$$

with $p_0(\cdot)$ denoting the actual probability of transmission of a given state (1) and $p_0(\rho_0) + p_0(\rho_1) = 1$. In quantum communication [20], there is a bound minimizing the receiver error (its lower bound)

$$P_H \;=\; \min_{\{\Pi_0,\Pi_1\}} p[\Pi_0, \Pi_1] \tag{17}$$

known as the Helstrom bound [20] which, for a binary communication using the alphabet in Equation (10) reads [20,21]

$$P_H \;=\; \frac{1}{2}\left(1 - \sqrt{1 - 4p_0(\rho_0)p_0(\rho_1)|\langle 0|J,0\rangle|^2}\right). \tag{18}$$

For the particular class of the Gazeau–Klauder coherent states studied in this paper, the Helstrom bound can be calculated explicitly:

$$P_H \;=\; \frac{1}{2} - \frac{1}{2}\sqrt{1 - \left(\left|I_{\frac{1-\mu}{\mu}}\left(2\sqrt{\frac{J}{\mu}}\right)\Gamma\left(\mu^{-1}\right)\left(\frac{J}{\mu}\right)^{1/2\,\frac{-1+\mu}{\mu}}\right|\right)^{-1}}. \tag{19}$$

The Helstrom bound (19) for a binary communication with Gazeau–Klauder coherent states as letters of an alphabet for different values of μ in Equation (5) is presented in Figure 2.

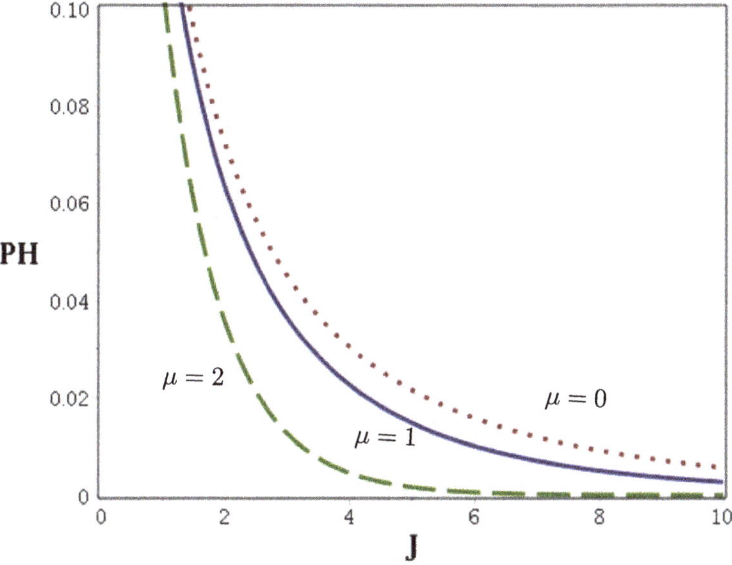

Figure 2. Helstrom bound P_H given by Equation (19) depicted for selected values of μ. For the sake of clarity, the range of P_H in the figure is limited to $P_H \leq 0.1$.

Let us notice that for small values of $J \leq 3$ the Helstrom bound P_H remains almost unaffected by nonlinearity even with relatively large amplitude $\mu \approx 2$. With increasing J the effect of nonlinearity becomes more apparent resulting in lowering the value of the Helstrom bound, i.e., resulting in an advantageous smaller minimal probability of receiver error.

3. Discussion

Quantum communication implemented with quantum optical states and devices seems to be one of the most promising for future developments. Non-randomized coherent states are natural candidates for letters of an alphabet used in communication [2–4]. Creation and manipulation of such states do not require devices with a 'hard' nonlinearity. In this work we studied a binary alphabet consisting of two Gazeau–Klauder coherent states [8] as an alternative for a well studied choice of standard

Perelomov coherent states. We utilize Gazeau–Klauder states calculated for a simplest nonlinear Kerr-type medium Equation (5) and we provide explicit analytic formulas for trace distance between Gazeau–Klauder and Perelomov states serving as codewords of the two alternative binary alphabets.

Despite that Gazeau–Klauder generalized coherent states [8] (used instead of the standard Perelomov) are harder in production [22], they can be, as we showed in this work, advantageous. At the cost of coping with a relatively well known Kerr-type nonlinear bosonic oscillator, present also beyond typical optical context [23], one gets a communication scheme with a smaller value of the Helstrom bound.

As coherent states with a non-randomized phase have recently attracted new attention [5] we believe that our analysis, despite its simplicity, can serve as a modest theoretical contribution for further practical developments utilizing Gazeau–Klauder coherent states in quantum communication and information processing and in a context of hybrid protocols [24].

4. Materials and Methods

Coherent states technique, quantum detection theory, quantum information with quantum optical implementations of quantum communication [4,6,8].

Author Contributions: Conceptualization, J.D. and J.Ł.; methodology, J.D.; software, J.D.; formal analysis, J.Ł.; writing—original draft preparation, J.D.; writing—review and editing, J.Ł. All authors have read and agree to the published version of the manuscript.

Funding: This work has been supported by the NCN Grant 2015/19/B/ST2/02856.

Conflicts of Interest: The authors declare no conflict of interest.

References

1. Akira Furusawa, A.; van Loock, P. *Quantum Teleportation and Entanglement: A Hybrid Approach to Optical Quantum Information Processing*; Wiley: Hoboken, NJ, USA, 2011.
2. Lo, H.K.; Preskill, J. Security of Quantum Key Distribution Using Weak Coherent States with Nonrandom Phases. *Quantum Inf. Comput.* **2007**, *7*, 431–458.
3. Ralph, T.C.; Gilchrist, A.; Milburn, G.J.; Munro, W.J.; Glancy, S. Quantum computation with optical coherent states. *Phys. Rev. A* **2003**, *68*, 042319. [CrossRef]
4. Gazeau, J.P. Coherent states in Quantum Information: An example of experimental manipulations. *J. Phys. Conf. Ser.* **2010**, *213*, 012013. [CrossRef]
5. Liu, L.; Wang, Y.; Lavie, E.; Wang, C.; Ricou, A.; Guo, F.Z.; Lim, C.C.W. Practical Quantum Key Distribution with Non-Phase-Randomized Coherent States. *Phys. Rev. Appl.* **2019**, *12*, 024048. [CrossRef]
6. Perelomov, A. *Generalized Coherent States and Their Applications*; Springer: Berlin/Heidelberg, Germany, 1986.
7. Gottesman, D.; Kitaev, A.; Preskill, J. Encoding a qubit in an oscillator. *Phys. Rev. A* **2001**, *64*, 012310. [CrossRef]
8. Gazeau, J.P.; Klauder, J.R. Coherent states for systems with discrete and continuous spectrum. *J. Phys. A Math. Gen.* **1999**, *32*, 123–132. [CrossRef]
9. Hollingworth, J.M.; Konstadopoulou, A.; Chountasis, S.; Vourdas, A.; Backhouse, N.B. Gazeau-Klauder coherent states in one-mode systems with periodic potential. *J. Phys. A Math. Gen.* **2001**, *34*, 9463–9474. [CrossRef]
10. Antoine, J.P.; Gazeau, J.P.; Monceau, P.; Klauder, J.R.; Penson, K.A. Temporally stable coherent states for infinite well and Pöschl–Teller potentials. *J. Math. Phys.* **2001**, *42*, 2349–2387. [CrossRef]
11. Roy, B.; Roy, P. Gazeau–Klauder coherent state for the Morse potential and some of its properties. *Phys. Lett. A* **2002**, *296*, 187–191. [CrossRef]
12. Roy, P. Quantum statistical properties of Gazeau–Klauder coherent state of the anharmonic oscillator. *Opt. Commun.* **2003**, *221*, 145–152. [CrossRef]
13. Novaes, M.; de Aguiar, M.A.M.; Hornos, J.E.M. Generalized coherent states for the double-well potential. *J. Phys. A Math. Gen.* **2003**, *36*, 5773–5786. [CrossRef]

14. Popov, D.; Sajfert, V.; Zaharie, I. Pseudoharmonic oscillator and their associated Gazeau–Klauder coherent states. *Phys. A Stat. Mech. Appl.* **2008**, *387*, 4459–4474. [CrossRef]
15. Dajka, J.; Łuczka, J. Gazeau–Klauder cat states. *J. Phys. A Math. Theor.* **2012**, *45*, 244006. [CrossRef]
16. Ching, C.L.; Ng, W.K. Deformed Gazeau-Klauder Schrödinger cat states with modified commutation relations. *Phys. Rev. D* **2019**, *100*, 085018. [CrossRef]
17. Kitagawa, M.; Yamamoto, Y. Number-phase minimum-uncertainty state with reduced number uncertainty in a Kerr nonlinear interferometer. *Phys. Rev. A* **1986**, *34*, 3974–3988. [CrossRef] [PubMed]
18. Milburn, G.J. Quantum and classical Liouville dynamics of the anharmonic oscillator. *Phys. Rev. A* **1986**, *33*, 674–685. [CrossRef] [PubMed]
19. Nielsen, M.A.; Chuang, I.L. *Quantum Computation and Quantum Information*; Cambridge University Press: Cambridge, UK, 2010.
20. Helstrom, C.W. *Quantum Detection and Estimation Theory*; Academic Press: Cambridge, MA, USA, 1976.
21. Helstrom, C.W. Quantum detection and estimation theory. *J. Stat. Phys.* **1969**, *1*, 231–252. [CrossRef]
22. Yadollahi, F.; Tavassoly, M. A theoretical scheme for generation of Gazeau–Klauder coherent states via intensity-dependent degenerate Raman interaction. *Opt. Commun.* **2011**, *284*, 608–612. [CrossRef]
23. Aldana, S.; Bruder, C.; Nunnenkamp, A. Equationivalence between an optomechanical system and a Kerr medium. *Phys. Rev. A* **2013**, *88*, 043826. [CrossRef]
24. Iliyasu, A.M.; Venegas-Andraca, S.E.; Yan, F.; Sayed, A. Hybrid Quantum-Classical Protocol for Storage and Retrieval of Discrete-Valued Information. *Entropy* **2014**, *16*, 3537–3551. [CrossRef]

Article

Electromagnetically Induced Transparency in Media with Rydberg Excitons 1: Slow Light

David Ziemkiewicz

Institute of Mathematics and Physics, UTP University of Science and Technology, Al. Prof. S. Kaliskiego 7, 85-789 Bydgoszcz, Poland; david.ziemkiewicz@utp.edu.pl

Received: 19 December 2019; Accepted: 1 February 2020; Published: 4 February 2020

Abstract: In this paper, we show that Electromagnetically Induced Transparency (EIT) can be realized in mediums with Rydberg excitons. With realistic, reliable parameters which show good agreement with optical and electro-optical experiments, as well as the proper choice of Rydberg exciton states in the Cu_2O crystal, we indicate how the EIT can be performed. The calculations show that, due to a large group index, one can expect the slowing down of a light pulse by a factor of about 10^4 in this medium.

Keywords: electromagnetically induced transparency

1. Introduction

In recent times, a lot of attention has been directed at the subject of excitons in bulk crystals due to the experimental observation of the so-called yellow exciton series in Cu_2O, up to a large principal quantum number of $n = 25$ [1–4]. In analogy to atomic physics, such excitons have been named Rydberg excitons (RE). By virtue of their specific properties, Rydberg excitons are of great interest in solid-state and optical physics. These objects, whose size scales as n^2 and where the energy spacing of neighbouring states decreases as n^{-3}, are well-suited for performing experiments in parameter ranges that are quite different from other quantum systems, such as atomic vapours. The small binding energy of RE makes them more sensitive to external magnetic and electric fields than other systems. The observation and detailed description of Rydberg excitons has opened a new field in condensed matter spectroscopy. Their specific properties have motivated both theoretical and experimental interest in this field, with studies ranging from their spectroscopy—that is, optical [2], electro-optical [3], and magneto-optical spectra [4,5] through non-atomic scaling laws [6] to quantum chaos [7].

In the last two years, particular interest has been directed to the study of the dynamical properties of systems with RE. Due to a large orbital radius, their dipole moments are exceptionally large for higher values of n, so that the interaction between higher Rydberg excitons is exceptionally strong and leads to the so-called Rydberg blockade. The appearance of a Rydberg blockade separates linear and nonlinear regimes of dynamical phenomena with RE. In the nonlinear regime, the possibility of observing giant optical nonlinearities of RE in microcavities has been considered [8], while the paper [9] deals with strong interaction phenomena at very low densities of RE, which enables one to determine the contribution of the nonlinear optical response of the medium. In the case of a smaller exciton density, it is possible to remain in the linear range, for which a single-photon source has been proposed [10]. RE mediums could be used as a gain system for solid-state, highly tunable masers, allowing to achieve output powers of order ranging from 10^{-6} to 10^{-2} W in continuous and pulse mode, respectively [11]. The realization of light slowing or storing and retrieving experiments in crystal with RE could unlock a plethora of dynamical effects, which might be observed in such media. These phenomena are based on the electromagnetically induced transparency (EIT), which has various potential applications in both classical and quantum optics. Słowik et al. [12] theoretically proposed to implement single-qubit gates and a two-qubit controlled not (CNOT) gate operating on polarized photons based on light storage. Recently, a highly efficient quantum memory protocol based on EIT has been experimentally demonstrated for cold rubidium [13] and cesium atoms [14]. Experimental evidence of spinor slow light as a quantum memory for two-color qubits in cold rubidium has also been presented [15].

EIT [16] is one of the most important effects utilized in quantum optics as it allows for the coherent control of materials' optical properties. The generic EIT utilizes extraordinary dispersive properties of an atomic medium with three active states in the Λ configuration. This phenomenon leads to a significant reduction of absorption of a weak, resonant laser field (probe field) by irradiating the medium with a strong control field, and thus making an otherwise opaque medium transparent. This leads to dramatic changes of dispersion properties of the system. Absorption forms a dip called the transparency window and approaches zero, while the dispersion in the vicinity of this region becomes normal with a steep slope, which increases for a decreasing control field—the resonant probe beam is transmitted almost without losses. It should be stressed that there is a significant distinction between Electromagnetically Induced Transparency and Autler–Townes Splitting (ATS) [17]. The Autler–Townes effect results in a "split" transition within a coupled three-level system due to the AC Stark effect and emerges in the strong coupling limit of the more recently discovered electromagnetically induced transparency [17]. The EIT effect is described by the formation of a dark state due to a destructive quantum interference between the transition pathways. Both ATS and EIT result in a transparency feature, which is qualitatively identified as a wide spectral region between the split-absorption peaks for ATS, but a narrow transmission window within a single-absorption peak for EIT. This common feature has been at the centre of long-standing confusion as to whether an observed transparency is due to EIT or ATS, and as such, the distinction between the two is an active topic of research in the quantum optics community. It has recently been shown theoretically [18], as well as experimentally [19] that it is possible to objectively distinguish between the regime where EIT dominates (EIT regime) and the one where ATS dominates (ATS regime). Usually, EIT is explained by the destructive quantum interference between different excitation pathways of the excited state, or alternatively, in terms of a dark superposition of states. For at least 20 years, there has been a considerable level of activity devoted to the EIT [20], which was motivated by the recognition of a number of its potential applications, such as slowing and storing the light (see e.g., [21–23] and the References therein). Some of the most exciting developments on EIT refer to their manifestations in artificial atoms [24–27]. A remarkable quenching of absorption due to EIT in an undoped bulk crystal of Cu_2O in a Λ configuration involving lower levels was examined in [28], and evidence of quantum coherences has been pointed out to exist in Cu_2O [29].

In Rydberg systems, a ladder configuration enables one to couple long-living metastable, initially empty upper levels with the lower levels coupled by the probe field. Demonstration of EIT in Rydberg atoms involving the ladder levels scheme by Mohapatra et al. [30] has taken advantage of their unique properties and entailed investigations concerning transient and steady-state EIT spectra in Rydberg systems [31]. The Rydberg excitons offer an unprecedented potential to study the above-mentioned phenomena in solids—excitons in Cu_2O offer a great variety of accessible states for creating the ladder configuration, which enables such a choice of coupling that could be realized by accessible lasers, or which could eventually to be suitable for desirable coherent interaction implementations. The first step toward the study of the photon blockade, many-body physics with Rydberg excitons and quantum non-linear optics is the realization of Rydberg EIT. The optical nonlinearity can be significantly enhanced in the presence of quantum interference in EIT systems, so examination of this phenomena in RE may be the first step to proposing an eventual application of this medium in quantum information processing.

Here, we focus our interest on the dynamical aspects and properties of Rydberg excitons in Cu_2O, and we show that the Rydberg excitonic states can be used to realize the EIT. We indicate excitonic states which guarantee the most efficient realization of the experiment, taking into account our previous results concerning excitonic resonances [2–5], as well as damping parameters and the matrix elements of the interband dipole operator. The influence of nonlinear interactions will also be taken into account. Moreover, a study of Rydberg EIT spectra will be performed in the presence of dipolar interactions, and their dependence on excitonic density, principal quantum numbers of Rydberg states, and control beam intensities will be discussed.

Our paper is organized as follows. In Section 2 we present the assumptions of the considered model and solve the time evolution equations, obtaining an analytical expression for the susceptibility and the group index. We use the obtained expression to compute the real and imaginary part of the susceptibility and the group index (Section 3) for a Cu_2O crystal slab. We examine in detail the influence of the control field on the dispersion, absorption, and the group index of the medium. In Section 4, we draw conclusions based on the model studied in this paper and indicate the optimal choice of Rydberg states to realize the EIT and light slowing in the linear regime.

2. Theory

The phenomenon of EIT, described qualitatively above, has been studied theoretically in various configurations of the transitions, probe, and control beams. Below, we propose a theoretical description for the case when the atomic transitions are replaced by intra-excitonic transitions in Rydberg excitons. The condensed matter exhibits quite a variety of three-level systems where an induced transparency could be achieved in much the same way as done in atomic mediums. Yet, dephasings, which can easily break the coherence of the population trapping state, are typically much faster in solids than in atomic vapours; it has caused great difficulty in observing a large, electromagnetically induced transparency effect in solids. We believe that this difficulty could be overcome by using the Rydberg exciton states. In particular, the higher states have exceptionally long lifetimes (which are proportional to n^2 [1]) and the dephasing can reach values which enables observation of the EIT effect.

Below, we consider a Cu_2O crystal as a medium where the EIT phenomenon can be realized. We use a ladder configuration (Figure 1) consisting of three levels, a, b, and c. As in previous works on Rydberg excitons, we focus our attention on the so-called yellow series associated with the lowest inter-band transition between the Γ_7^+ valence band and the Γ_6^+ conduction band. Because both band-edge states are of even parity, the lowest $1S$ exciton state is dipole-forbidden, whereas all the P states are dipole-allowed; the $1S$ to nP transition is also allowed. We have chosen the valence band as the b state. As a practical example, the n_1P and n_2S excitonic states are chosen. The proposed ladder system contains only dipole-allowed $S \rightarrow P$ transitions.

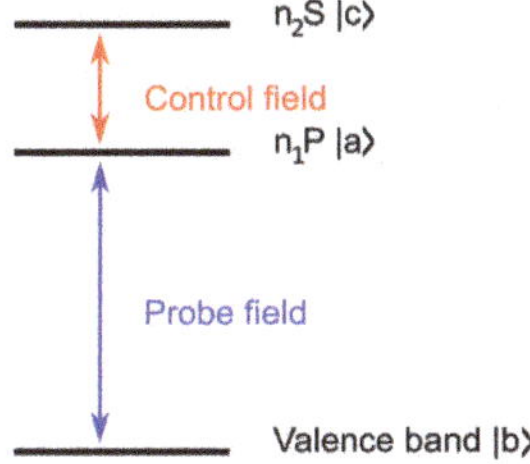

Figure 1. Schematic of the considered ladder EIT system.

Let the probe/signal field of frequency ω_1 and amplitude ε_1 couple the ground state b of energy E_b with an excited state a of energy E_a. The control field of frequency ω_2 and amplitude ε_2 couples the state c of energy E_c with the state a, as it is illustrated in Figure 1. The Hamiltonian of such a three-level system interacting with an electromagnetic wave in the rotating wave approximation leads to

$$
\begin{aligned}
H \;=\; & E_a|a\rangle\langle a| + E_b|b\rangle\langle b| + E_c|c\rangle\langle c| \\
& + \{-\hbar\Omega_1(z,t)\exp[-i(\omega_1 t - k_1 z)]|a\rangle\langle b| \\
& -\hbar\Omega_2(z,t)\exp[-i(\omega_2 t - k_2 z)]|a\rangle\langle c| + h.c.\} + V_{nl},
\end{aligned}
\tag{1}
$$

where h.c. stands for hermitian conjugate. The k_1, k_2, $\Omega_1(z,t) = (1/\hbar)d_{ab}\varepsilon_1(z,t)$, $\Omega_2(z,t) = (1/\hbar)d_{ac}\varepsilon_2(z,t)$ are the wave vectors and real Rabi frequencies corresponding to the particular couplings, respectively. The d_{ij} are the dipole transition moments related to the specific transitions, and V_{nl} describes the nonlinear interactions between Rydberg excitons. One of the characteristic features of Rydberg media is the existence of the Rydberg blockade. Strong dipolar interaction between Rydberg excitons, which strongly depends on their distance, shifts the Rydberg levels, preventing the optical excitation of nearby excitons by shifting their corresponding levels out of the resonance with the exciting electromagnetic field. The term Rydberg blockade refers to the case where the interaction-induced shift is much larger than the EIT linewidth. In this case, the resulting nonlinearity can be interpreted as a switch from three-level EIT susceptibility to two-level susceptibility [32]. Resonant absorption and exciton creation are no longer possible inside the blockade volume V_{bloc}, in which dipole interaction energy is larger than the EIT linewidth. This effect influences the generic, linear EIT, so that the nonlinear modification should be taken into account [32]. In the presented manuscript, we consider the situation when one remains in the linear regime, that is, for exciton density small enough to avoid the Rydberg blockade. The presented calculations allow one to make such a choice of excitonic states where linear approximation still holds, such as the blockade-induced shift $\delta_{Rydb}/\Gamma_{ab} << 1$; this is ensured by the exciton density which is two to three orders of magnitude below the limit where the excitons are packed closely together and $\delta_{Rydb} \sim \Gamma_{ab}$.

The time evolution of the system is governed by the von Neumann equation with a phenomenological relaxation contribution

$$
i\hbar\frac{d\sigma}{dt} = [H,\sigma] + R\sigma,
\tag{2}
$$

where $\sigma(z,t)$ denotes the density matrix for an exciton at position z and time t, and R is the relaxation operator accounting for all relaxation processes in the medium.

By neglecting propagation effects for the control field and denoting the probe and control beam detunings by $\delta_1 = (E_a - E_b)/\hbar - \omega_1$ and $\delta_2 = (E_c - E_a)/\hbar - \omega_2$, respectively, the evolution of the system can be described by the following equations:

$$
\begin{aligned}
i\dot{\sigma}_{aa} \;=\; & -\Omega_1(\sigma_{ba} - \sigma_{ab}) - \Omega_2(\sigma_{ca} - \sigma_{ac}) - i\Gamma_{ab}\sigma_{aa} + i\Gamma_{ca}\sigma_{cc}, \\
i\dot{\sigma}_{bb} \;=\; & \Omega_1(\sigma_{ba} - \sigma_{ab}) + i\Gamma_{ab}\sigma_{aa} + i\Gamma_{cb}\sigma_{cc}, \\
i\dot{\sigma}_{cc} \;=\; & -\Omega_2(\sigma_{ac} - \sigma_{ca}) - i\Gamma_{ca}\sigma_{cc} - i\Gamma_{cb}\sigma_{cc}, \\
i\dot{\sigma}_{ab} \;=\; & (\delta_1 - i\gamma_{ab})\sigma_{ab} - \Omega_1(\sigma_{bb} - \sigma_{aa}) - \Omega_2\sigma_{cb}, \\
i\dot{\sigma}_{bc} \;=\; & (\delta_2 - \delta_1 - i\gamma_{cb})\sigma_{bc} + \Omega_2\sigma_{ba} - \Omega_1\sigma_{ac} + \sum_{j\neq i} V_{ij}\sigma^j_{cc}\sigma^i_{bc}, \\
i\dot{\sigma}_{ca} \;=\; & (\delta_2 - i\gamma_{ca})\sigma_{ca} - \Omega_2(\sigma_{cc} - \sigma_{aa}) - \Omega_1\sigma_{cb} + \sum_{j\neq i} V_{ij}\sigma^j_{cc}\sigma^i_{ca}..
\end{aligned}
\tag{3}
$$

At large separations between RE, the dipole–dipole interaction is dominant, and the potential V_{ij} strongly depends on the separation distance between Rydberg excitons [1]. The parameters Γ_{ij}, $i \neq j$ describe the damping of exciton states and are determined by temperature-dependent homogeneous broadening due to phonons, and broadening due to structural imperfections and eventual impurities.

The relaxation damping rates for the coherence are denoted by $\gamma_{ij} \approx \Gamma_{ij}/2$, $i \neq j$ [28]. It should be noted that in the above equations, only the relaxations inside the three level system are considered, so that the total probability for the populations of the three levels is conserved: $\sigma_{aa} + \sigma_{bb} + \sigma_{cc} = 1$.

Due to the fact that at $t = 0$, $\sigma_{bb}(t = 0) = 1$ for a weak probe field (i.e., $|\Omega_1|^2 << |\Omega_2|^2$), in the first-order perturbation with respect to the probe field, the evolution of our system, given by Bloch equations, reduces to a set of the following equations for the density matrix:

$$i\dot{\sigma}_{ab} = (\delta_1 - i\gamma_{ab})\sigma_{ab} - \Omega_1 - \Omega_2\sigma_{cb}, \tag{4}$$

$$i\dot{\sigma}_{bc} = (\delta_2 - \delta_1 - i\gamma_{cb})\sigma_{bc} + \Omega_2\sigma_{ba} + \sum_{j \neq i} V_{ij}\sigma_{cc}^j\sigma_{bc}^i. \tag{5}$$

Taking into account the fact that for a weak probe pulse, for a given frequency, the polarization of the medium P_1 is proportional to the signal field ε and susceptibility χ, it takes the form

$$P_1 = \epsilon_0\chi(\omega_1)\varepsilon_1(\omega_1) = Nd_{ba}\sigma_{ab}, \tag{6}$$

where N is the density of excitons, d_{ba} is the transition dipole matrix element [11], and ϵ_0 is the vacuum dielectric permittivity. In the limit of the low probe intensity, the susceptibility, which follows from the steady-state solutions of Equations (4) to (5), can be expanded into the Taylor series

$$\chi(\omega_1) = \chi^{(1)} + \chi^{(3)}\Omega_1^2 = -\frac{N|d_{ab}|^2}{\hbar\epsilon_0}\frac{1}{\omega_1 - \delta_1 + i\gamma_{ab} - \frac{|\Omega_2|^2}{\omega_1 - \delta_1 + \delta_2 + i\gamma_{cb}}} + \chi^{(3)}\Omega_1^2, \tag{7}$$

where $\chi^{(1)}$ is the linear, and $\chi^{(3)}$ is the nonlinear part of electric susceptibility. The formula for the linear part $\chi^{(1)}$ follows from the steady-state solution of Equations (4) to (5) and $\chi^{(3)}$ is treated as a correction, which depends on interaction potential V_{ij} and causes the nonlinear modification of EIT due to the strong interactions between Rydberg excitons. For a large exciton separation R, this potential has the form C_6/R^6, with C_6 being a coefficient proportional to n^{11}, which drastically increases for higher Rydberg excitons. On the other hand, for closer separation of excitons, when the dipole–dipole interaction becomes comparable or even exceeds the energy spacing between excitonic levels, this interaction has the Foerster type and is proportional to n^4/R^3 [1]. In our system, we remain in the low-density regime where the C_6/R^6 interaction is dominant.

The susceptibility is a complex, rapidly varying function of ω_1, with its real part responsible for the dispersion and the imaginary part describing the absorption. If the excitons are driven on a single-photon resonance $\delta_1 \approx 0$, the main nonlinear effect will be the nonlinear absorption. Neglecting the transverse probe beam dynamics for lower excitonc states, that is, when van der Waals is dominant, it is possible to analytically calculate this nonlinear part of susceptibility [33],

$$\chi_{Re}^{(3)} = \frac{4\sqrt{2}\pi^3\gamma_{ab}\Omega_2^4 C_6|C_6|^{-1/2}}{k^3\sqrt{\gamma_{ab}}(\gamma_{ca}\gamma_{ab} + \Omega_2^2)^{7/2}}N^2; \quad \chi_{Im}^{(3)} = |\chi_{Re}^{(3)}|. \tag{8}$$

It should be stressed that the interaction between Rydberg excitons modifies the system in a nonlinear way by shifting the Rydberg levels, effectively changing the control field detuning and influencing the absorption and the width of the transparency window. Note that the equality of real and imaginary parts of $\chi^{(3)}$ holds in a limited frequency range around the $a \to b$ resonance [33].

Due to the dependence of the refractive index $n(\omega_1) = \sqrt{\epsilon_b + \operatorname{Re}\chi(\omega_1)}$ on the medium dispersion, we define the group index

$$n_g(\omega_1) = c/v_g = 1 + \frac{1}{2}\operatorname{Re}\chi(0) + \frac{\omega_1}{2}\frac{\partial}{\partial\omega_1}\operatorname{Re}\chi(\omega_1), \tag{9}$$

which accounts for the time delay of a pulse propagating in a medium with the group velocity

$$v_g = \frac{c}{1 + \frac{1}{2}\mathrm{Re}\,\chi(0) + \frac{\omega_1}{2}\frac{\partial\mathrm{Re}\,\chi(0)}{\partial\omega_1}}. \tag{10}$$

The slope of dispersion relation inside the transparency window determines the velocity of the pulse propagating inside the medium. Note that the derivative of the real part of the susceptibility may be positive (normal dispersion) or negative (anomalous dispersion); in the latter case, the group velocity may even become negative.

3. Numerical Results

We considered a Cu_2O crystal slab with a thickness of 30 μm as a medium where the EIT phenomenon can be realized. We used the ladder configuration (Figure 1), consisting of three levels. Let the probe/signal field of frequency ω_1 couple the ground state (valence band) with the excited state $n_1 P$. The control field of a much lower frequency ω_2 couples states $n_1 P$ and $n_2 S$. Using the Formulas (7) and (9), we calculate the real and imaginary part of the susceptibility and the group index. For the state energies and damping parameters, we use the values obtained from fitting to experimental data, as described in [3]. The transition dipole moments were calculated from hydrogen-like functions, as described in detail in [11]. In particular, we used Hydrogen P-orbital with the radius proportional to n^2 [1], which results in dipole moments in the range of 1.5 ea_b (1S state) up to over 150 ea_b (10P state) [11]. Note that instead of d_{ba}, we used the function $M_{01}(\mathbf{r})$, which is an analogue of the dipole moment element, but takes into account the smeared-out transition dipole density characteristic for coupling between the valence band and the $2P$ excitonic state [4]. Application of this smeared-out dipole density is justified by a good agreement with an experiment of our previous results for electro-optical properties of Rydberg excitons [3]. We chose the exciton density $N = 2.5 \cdot 10^{14}$ cm^{-3}, which is two orders of magnitude below the maximum value of $N \sim \frac{3\cdot10^{18}}{n_1^7} \approx 2.6 \cdot 10^{16}$ cm^{-3} for $n_1 = 2$ [1].

The system parameters are summarized in Table 1. We assumed that $\gamma_{ij} = \Gamma_{ij}/2$ [28]; the damping rates are calculated for $T = 10$ K, where the contribution of exciton scattering with acoustic phonons is relatively small, and the main factor is spontaneous, with an emission rate proportional to n^3 [11].

Table 1. Parameters used in calculations.

Parameter	
ω_1	3291 THz
ω_2	31.71 THz
γ_{ab}	4.558 THz
γ_{bc}	36.46 GHz
Γ_{ab}	9.116 THz
Γ_{bc}	72.92 GHz
Ω_2	20 GHz
N	$2.5 \cdot 10^{14}$ cm^{-3}
M_{01}	$2.1 \cdot 10^{-30}$ C·m

The choice of states n_1 and n_2 has a fundamental effect on the properties of the system. The interaction coefficient $C_6 \sim n_1^{11}$ shows the strongest dependence on the state number and puts an upper limit on the lower state n_1. Moreover, the maximum exciton density is inversely proportional to the Rydberg blockade volume $V_{bloc} \sim n_1^7$ (note that the upper state n_2 is empty, so that the blockade does not apply to it). The state n_2 affects the susceptibility by relaxation rates. All these factors indicate that a successful light slowdown can be achieved only for some parameters and the state configurations, where the linear part of susceptibility is dominating, and thus the group velocity is well-defined. Therefore, as a first step, we have investigated the range of states and control field intensities where the linear part of susceptibility is dominating, as shown in the Figure 2. As pointed out above, the nonlinearity is very sensitive to the state number and varies by many orders of magnitude. One can also see that, in general, linear response is obtained for weaker control fields, which is advantageous for a large slowdown factor, but results in a relatively narrow transparency window. In the range where $\chi^{(3)}$ is negligible, there is very little absorption inside the transparency window, and the group velocity V_g is a well-defined quantity that can be calculated in a linear regime from Equation (11).

The linear region is outlined on the insets of the Figure 2. One can see that, practically, only the lower states $n_1=1,\ldots4$ are usable, except for extremely small values of control field intensity which are impractical due to the fact that in EIT, the control field needs to be much stronger than the probe field, that is, $\Omega_2 >> \Omega_1$. On the other hand, the n_2 should be large; due to the fact that this level remains empty, it is not constrained by the Rydberg blockade.

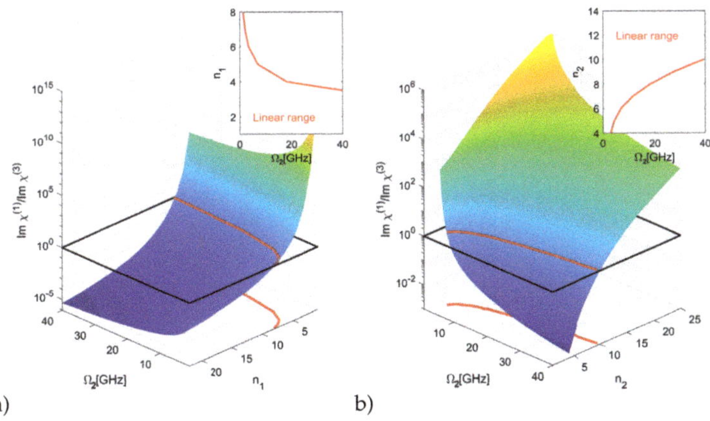

a)   b)

Figure 2. The ratio of the linear to nonlinear susceptibility as a function of control field strength and state number (**a**) n_1 and (**b**) n_2; colors added for clarity. Inset: the range of parameters where the linear susceptibility is larger.

The Figure 3 shows the medium susceptibility and the group index $n_g = c/V_g$. For a non-zero Rabi frequency Ω_2 of the control field, the imaginary part of the system's susceptibility reveals a dip in the Lorentzian absorption profile, called a transparency window. This means that the resonant probe beam which otherwise would be strongly absorbed is now transmitted almost without losses. The width of the transparency window is proportional to the square of the control field amplitude, and therefore, by increasing the control field strength, it is possible to open it out. This is shown in the Figure 4, where one can observe a dip in absorption (Im χ) around the zero-detuning part of the spectrum, which is accompanied by a very steep slope of the Re χ. That strong, normal dispersion inside the window is responsible for the reduction of the group velocity. The absorption at the resonance does not reach zero due to the finite value of the relaxation rate γ_{cb}, but it is indeed very small (absorption coefficient $\alpha \sim 3$ cm^{-1}). This is easily visible in the Figure 3, where the control field is set to $\Omega_2 = 10$ GHz. This means that the medium has become almost transparent for a probe pulse which now travels with a reduced velocity. Indeed, the group index shows that the impulse slowdown factor of about 3700 is expected. It should be stressed that the value of n_g is directly proportional to the magnitude of susceptibility changes, and thus, to the exciton density. Clearly, high values are favourable; for the lowest state, the densities on the order of 10^{19} cm^{-3} are possible [34,35]. In our calculations, we have chosen $N = 2.5 \times 10^{14}$ cm^{-3}, which corresponds to susceptibility values on the order of 0.03 (Figure 3), comparable with values presented in [28] for the low excitonic states and well below the limit imposed by the Rydberg blockade for the chosen state. It should be stressed that while these densities are orders of magnitude greater than in atomic EIT systems, the possible impulse slowdown is limited by much larger dissipation rates, particularly γ_{ab}.

As mentioned previously, there are many Rydberg state combinations that provide a working EIT system. The maximum possible slowdown in all sets of states $n_1 = 1, \ldots 10$, $n_2 = 2, \ldots 20$, a strong control field $\Omega_2 = 30$ GHz, and probe field $\Omega_1 = 3$ GHz is shown in the Figure 5a. The blue dots mark the systems where the linear part of susceptibility is much greater than the nonlinear, so that the group velocity is a well-defined quantity. The red dots mark the range where $Re\chi < Im\chi$, so that the calculated slowdown is only a rough estimation. Remarkably, a very large slowdown can be achieved for $n_1 = 1$ due to the small size of excitons and, correspondingly, large exciton density. In the strong control field regime, the linear systems can be based only on $n_1 = 1, 2, 3$; for higher states, the effect of a Rydberg blockade becomes significant. Moreover, it is beneficial to use the high upper state n_2 so that the probe and control field frequencies are similar, for example, $\omega_1 \approx \omega_2$. It should be pointed out that a high value of n_2 is only possible at highly cryogenic temperatures. Additionally, the probe beam may excite a group of closely-lying upper states, further degrading the performance [11].

By reducing the probe and control field strengths to $\Omega_1 = 0.3$ GHz and $\Omega_2 = 3$ GHz, correspondingly, one can extend the range of linear systems to $n_1 = 6$, as shown in the Figure 5b. In such a case, the transparency window is much narrower. Due to this, the slowdown in the systems with $n_1 < 5$ is limited by the large dissipation constant γ_{ab}, which prevents the transparency window from fully opening. On the other hand, the weaker probe and control fields result in a smaller density, so that the Rydberg blockade effect is delayed to $n_1 > 6$.

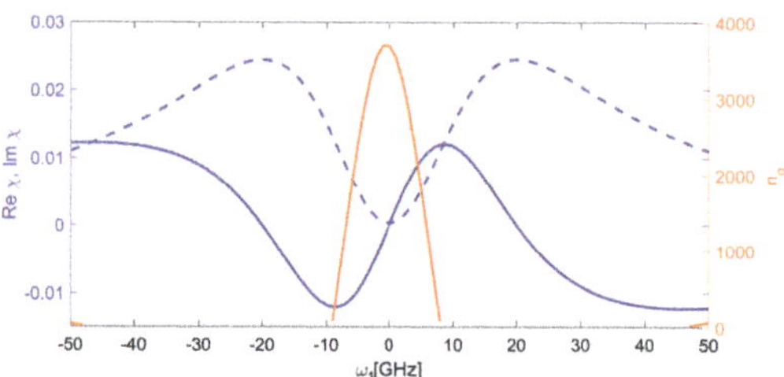

Figure 3. The real and imaginary parts of linear susceptibility and group index as a function of detuning, for $n_1 = 2$, $n_2 = 10$, $\Omega_2 = 10$ GHz, $\Omega_1 = 1$ GHz.

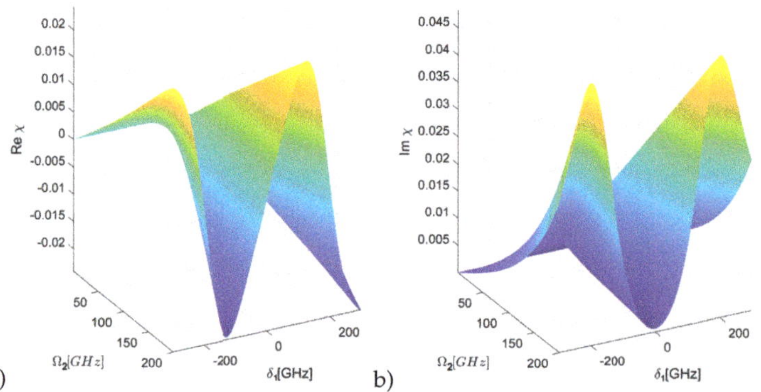

Figure 4. The real (**a**) and imaginary (**b**) part of the linear susceptibility as a function of control field strength and frequency for $n_1 = 2$, $n_2 = 10$, $\Omega_1 = 0.1\Omega_2$. Colors added for clarity.

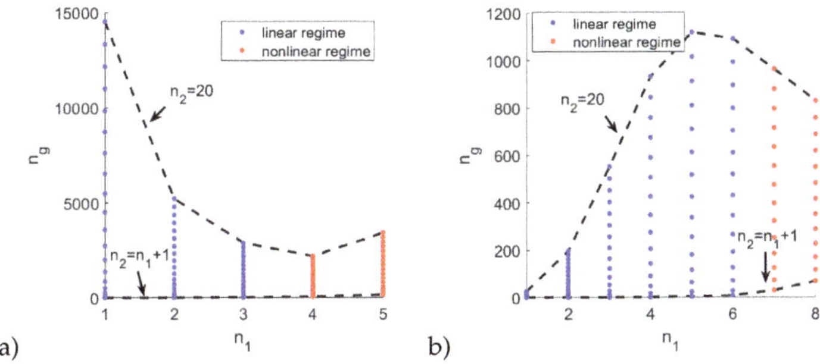

Figure 5. Maximum group velocity slowdown as a function of n_1 and n_2 for (**a**) $\Omega_2 = 30$ GHz (**b**) $\Omega_2 = 3$ GHz.

4. Conclusions

The Rydberg excitons are now hopefully emerging as a tool for quantum technology due to their unique properties, allowing for easy state selection and strong interaction with applied electromagnetic fields. Their unusual features can be useful for controlling matter–electromagnetic field interactions, which offers a new approach for studying semiconductor systems and also provides entirely new long-term perspectives for developing novel devices, which are more robust and compact than atomic systems. We have indicated the optimal states and well-justified parameters to attempt the observation of EIT in Rydberg excitons' Cu_2O media, which allows one to obtain a considerable value of the group index. Due to the coherence properties of Rydberg excitons, the manipulations of the medium transparency is possible; the width of the window and slowing down the group velocity of the pulse travelling inside the sample might be changed in a controlled way by the strength of the control field. The ability to control the group index on-demand enables one to store and retrieve light pulses, which is a first step toward quantum memory implementation. The method of precise dynamical control of the optical properties of the medium by optical means reveals new aspects of excitonic quantum optics and is supposed to lead to constructing an efficient tools for photonics, quantum switches, or multiplexers. Since the first observation of Rydberg excitons in 2014, their spectroscopic properties have mostly been investigated. Only a few observations have dealt with quantum optical aspects; the quantum coherence effects have been analyzed in [29], and a Rydberg exciton single-photon source has been proposed [10]. These studies have just opened up the possibility for a controlled light-matter coupling in a GHz regime with RE media. So far, the experimental demonstration of EIT in Rydberg excitons media has not been accomplished yet, but one may expect such experiments in the future. Performing the EIT in excitonic Rydberg media will be a step towards realization of controlled interaction of Rydberg excitons in integrated and scalable solid-state devices and potential implementation of this medium for quantum information processing.

Funding: Support from the National Science Centre, Poland (project OPUS 2017/25/B/ST3/00817) is greatly acknowledged.

Acknowledgments: I wish to thank Sylwia Zielińska - Raczyńska for a valuable discussion and an enthusiastic encouragement.

Conflicts of Interest: The author declare no conflict of interest.

References

1.   Kazimierczuk, T.; Fröhlich, D.; Scheel, S.; Stolz, H.; Bayer, M. Giant Rydberg excitons in the copper oxide Cu_2O. *Nature* **2014**, *514*, 343–347. [CrossRef]
2.   Zielińska-Raczyńska, S.; Czajkowski, G.; Ziemkiewicz, D. Optical properties of Rydberg excitons and polaritons. *Phys. Rev. B* **2016**, *93*, 075206. [CrossRef]

3. Zielińska-Raczyńska, S.; Ziemkiewicz, D.; Czajkowski, G. Electro-optical properties of Rydberg excitons. *Phys. Rev. B* **2016**, *94*, 045205. [CrossRef]

4. Zielińska-Raczyńska, S.; Ziemkiewicz, D.; Czajkowski, G. Magneto-optical properties of Rydberg excitons: Center-of-mass quantization approach. *Phys. Rev. B* **2017**, *95*, 075204. [CrossRef]

5. Zielińska-Raczyńska, S.; Fishman, D.A.; Faugeras, C.; Potemski, M.; van Loosdrecht, P.; Karpiński, K.; Czajkowski, G.; Ziemkiewicz, D. Magneto-excitons in Cu_2O: Theoretical model from weak to high magnetic fields. *New J. Phys.* **2019**, *21*, 103012. [CrossRef]

6. Heckötter, J.; Freitag, M.; Fröhlich, D.; Aßmann, M.; Bayer, M.; Semina, M.; Glazov, M. Scaling laws of Rydberg excitons. *Phys. Rev. B* **2017**, *96*, 125142. [CrossRef]

7. Aßmann, M.; Thewes, J.; Fröhlich, D.; Bayer, M. Quantum chaos and breaking of all anti-unitary symmetries in Rydberg excitons. *Nat. Mater.* **2016**, *15*, 741–745. [CrossRef]

8. Walther, V.; Johne, R.; Pohl, T. Giant optical nonlinearities from Rydberg excitons in semiconductor microcavities. *Nat. Commun.* **2018**, *9*, 1–6. [CrossRef]

9. Walther, V.; Krüger, S.O.; Scheel, S.; Pohl, T. Interactions between Rydberg excitons in Cu_2O. *Phys. Rev. B* **2018**, *98*, 1165201. [CrossRef]

10. Khazali, M.; Heshami, K.; Simon, C. Single-photon source based on Rydberg exciton blockade. *J. Phys. B* **2017**, *50*, 215301. [CrossRef]

11. Ziemkiewicz, D.; Zielińska-Raczyńska, S. Solid-state pulsed microwave emitter based on Rydberg excitons. *Opt. Express* **2019**, *27*, 16983–16994. [CrossRef] [PubMed]

12. Słowik, K.; Raczyński, A.; Zaremba, J.; Zielińska-Kaniasty, S. Light storage in a tripod medium as a basis for logical operations. *Opt. Commun.* **2012**, *285*, 2392–2396. [CrossRef]

13. Rastogi, A.; Saglamyurek, E.; Hrushevskyi, T.; Hubele, S.; LeBlanc, L.J. Discerning quantum memories based on electromagnetically-induced-transparency and Autler-Townes-splitting protocols. *Phys. Rev. A* **2019**, *100*, 012314. [CrossRef]

14. Vernaz-Gris, P.; Huang, K.; Cao, M.; Sheremet, A.S.; Laurat, J. Highly-efficient quantum memory for polarization qubits in a spatially-multiplexed cold atomic ensemble. *Nat. Commun.* **2018**, *9*, 1–6. [CrossRef]

15. Lee, M.-J.; Ruseckas, J.; Lee, C.-Y.; Kudriašov, V.; Chang, K.-F.; Cho, H.-W.; Juzeliānas, G.; Ite, A.Y. Experimental demonstration of spinor slow light. *Nat. Commun.* **2014**, *5*, 1–8. [CrossRef]

16. Harris, S.E. Electromagnetically Induced Transparency. *Phys. Today* **1997**, *50*, 36. [CrossRef]

17. Anisimov, P.M.; Dowling, J.P.; Sanders, B.C. Objectively discerning Autler-Townes splitting from electromagnetically induced transparency. *Phys. Rev. Lett.* **2011**, *107*, 163604. [CrossRef]

18. Peng, B.; Özdemir, S.; Chen, W.; Nori, F.; Yang, L. What is and what is not electromagnetically induced transparency in whispering-gallery microcavities. *Nat. Commun.* **2014**, *5*, 5082. [CrossRef]

19. Liu, Q.-C.; Li, T.-F.; Luo, X.-Q.; Zhao, H.; Xiong, W.; Zhang, Y.-S.; Chen, Z.; Liu, J.; Chen, W.; Nori, F. Method for identifying electromagnetically induced transparency in a tunable circuit quantum electrodynamics system. *Phys. Rev. A* **2016**, *93*, 053838. [CrossRef]

20. Fleischhauer, M.; Immamoglu, A.; Marangos, J.P. Electromagnetically induced transparency: Optics in coherent media. *Rev. Mod. Phys.* **2005**, *77*, 633. [CrossRef]

21. Raczyński, A.; Zaremba, J.; Zielińska-Kaniasty, S. Electromagnetically induced transparency and storing of a pair of pulses of light. *Phys. Rev. A* **2004**, *69*, 043801. [CrossRef]

22. Perdian, M.; Raczyński, A.; Zaremba, J.; Zielińska-Kaniasty, S. Controlling statistical properties of stored light. *Opt. Commun.* **2007**, *279*, 324–329.

23. Raczyński, A.; Zaremba, J.; Zielińska-Kaniasty, S. Beam splitting and Hong-Ou-Mandel interference for stored light. *Phys. Rev. A* **2007**, *75*, 013810. [CrossRef]

24. Ian, H.; Liu, Y.; Nori, F. Tunable electromagnetically induced transparency and absorption with dressed superconducting qubits. *Phys. Rev. A* **2010**, *81*, 063823. [CrossRef]

25. Sun, H.; Liu, Y.; Ian, H.; You, J.; Il'ichev, E.; Nori, F. Electromagnetically induced transparency and Autler–Townes splitting in superconducting flux quantum circuits. *Phys. Rev. A* **2014**, *89*, 063822. [CrossRef]

26. Wang, H.; Gu, X.; Liu, Y.-X.; Miranowicz, A.; Nori, F. Optomechanical analog of two-color electromagnetically induced transparency: Photon transmission through an optomechanical device with a two-level system. *Phys. Rev. A* **2014**, *90*, 023817. [CrossRef]

27. Jing, H.; Özdemir, S.; Geng, Z.; Zhang, J.; Lü, X.; Peng, B.; Yang, L.; Nori, F. Optomechanically-induced transparency in parity-time-symmetric microresonators. *Sci. Rep.* **2015**, *5*, 9663. [CrossRef]

28. Artoni, M.; La Rocca, G.; Bassani, F. Electromagnetic-induced transparency of Wannier-Mott excitons. *Europhys. Lett.* **2000**, *49*, 445. [CrossRef]

29. Grünwald, P.; Aßmann, M.; Heckötter, J.; Fröhlich, D.; Bayer, M.; Stolz, H.; Scheel, S. Signatures of Quantum Coherences in Rydberg Excitons. *Phys. Rev. Lett.* **2016**, *117*, 133003. [CrossRef]

30. Mohapatra, A.K.; Jackson, T.R.; Adams, C.S. Coherent Optical Detection of Highly Excited Rydberg States Using Electromagnetically Induced Transparency. *Phys. Rev. Lett.* **2007**, *98*, 113003. [CrossRef]

31. Zhang, Q.; Bai, Z.; Huang, G. Fast-responding property of electromagnetically induced transparency in Rydberg atoms. *Phys. Rev. A* **2018**, *97*, 043821. [CrossRef]

32. Firstenberg, O.; Adams, C.; Hofferberth, S. Nonlinear quantum optics mediated by Rydberg interactions. *J. Phys. B At. Mol. Opt. Phys.* **2016**, *49*, 152003. [CrossRef]

33. Sevinçli, S.; Henkel, N.; Ates, C.; Pohl, T. Nonlocal Nonlinear Optics in Cold Rydberg Gases. *Phys. Rev. Lett.* **2011**, *107*, 153001. [CrossRef] [PubMed]

34. Kavoulakis, G.M.; Baym, G.; Wolfe, J.P. Quantum saturation and condensation of excitons in Cu_2O: A theoretical study. *Phys. Rev. B* **1996**, *53*, 7227. [CrossRef]

35. Karpinska, K.; Mostovoy, M.; van der Vegte, M.; Revcolevschi, A.; van Loosdrecht, P.H.M. Decay and coherence of two-photon excited yellow orthoexcitons in Cu_2O. *Phys. Rev. B* **2005**, *72*, 155201. [CrossRef]

Article

Electromagnetically Induced Transparency in Media with Rydberg Excitons 2: Cross-Kerr Modulation

David Ziemkiewicz * and Sylwia Zielińska-Raczyńska

Institute of Mathematics and Physics, UTP University of Science and Technology, Al. Prof. S. Kaliskiego 7, 85-789 Bydgoszcz, Poland; sziel@utp.edu.pl
* Correspondence: david.ziemkiewicz@utp.edu.pl

Received: 19 December 2019; Accepted: 28 January 2020; Published: 30 January 2020

Abstract: By mapping photons into the sample of cuprous oxide with Rydberg excitons, it is possible to obtain a significant optical phase shift due to third-order cross-Kerr nonlinearities realized under the conditions of electromagnetically induced transparency. The optimum conditions for observation of the phase shift over π in Rydberg excitons media are examined. A discussion of the application of the cross-phase modulations in the field of all-optical quantum information processing in solid-state systems is presented.

Keywords: electromagnetically induced transparency; cross-Kerr nonlinearity

---

1. Introduction

The recent development of various light manipulation techniques arising from electromagnetically induced transparency (EIT) has made it possible to slow down the pulse (photons), store, and retrieve them, preserving the phase relations [1]. At the level of single photons, the EIT enables the quantum information carried by photons to be mapped in the form of quantum coherences inside the medium, effectively creating a quantum memory. It allows one to transfer quantum states between photons and matter. Recently, it has been shown that EIT significantly enhances the optical nonlinearity [2]. EIT has been proposed as a way to greatly enhance cross-phase modulation (XPM), which refers to the phenomenon where the phase of one photon is modulated by another photon [3]. One of the widely explored schemes to enhance cross-phase modulation is based on the Kerr-EIT-like interaction between two weak optical fields [4]. Recently, Bai et al. [5] considered strong Rydberg–Rydberg interactions which are the source of third- and even fifth-order Kerr nonlinearities. Interactions between photons realized by nonlinear optical mechanisms are essential to quantum information processing, quantum teleportation, and quantum logic gates [6–8]. Due to the large nonlinear susceptibilities at low light levels, the EIT-based XPM in atomic vapors makes single-photon operations feasible and can lead to applications in quantum information manipulation. XPM has been considered as a promising means of quantum communication and quantum computation. The large nonlinearity at the single-photon level could pave the way for the implementation of universal quantum gates. However, realizing large nonlinearity at such low light levels has been a great challenge for scientists in the past decade [9,10].

Solid bulk media are systems well worth considering for storing and processing quantum information because they have a number advantages over atomic gases, where many experiments have been done (for a recent review see Ref. [11]). They are easy to prepare, diffusion processes are not very fast, and much higher densities of interacting particles can be achieved [12]. One common class of solids used within the quantum information context are the rare-earth-metal-doped crystals, where a long time period of information storage has been achieved (i.e., over one minute) [13,14] Nitrogen-vacancy centers in diamond are also of interest [15], which have a relatively long spin coherence. Another class of solid-state systems where EIT occurs are so-called artificial atoms [16–19]. Recently, the Rydberg excitons (REs) have attracted a great deal of attention due to their exciting features: the distinct combination of their long radiative life-times, sensitivity to external fields, and strong dipolar interactions [20] could be exploited to realize quantum interfaces for quantum information processing. Rydberg excitonic samples smaller than other solid-state systems mentioned above. The observation of dipolar blockade in bulk Cu_2O [21], quantum coherence [22], and single-photon source based on RE blockade [23] were performed in samples of micrometer scale. The realization of these experiments has unlocked a plethora of dynamic effects which might be observed in Rydberg excitons media [24–27]. One example is the electromagnetically induced transparency discussed in our previous paper [28], the performance of which in Cu_2O bulk crystal will be the next step towards the potential implementation of this medium for quantum information processing. This paper follows up our previous work [28], where the optimal conditions for performing EIT in the linear regime were discussed. Here, by expanding these considerations and results, we propose to explore the scheme that enables one to induce a substantial nonlinear interaction and cross-phase modulation between two slow-light narrow pulses, realized in a cuprous oxide crystal with RE. This nonlinearity may be reached by disturbing the two-photon resonance condition in a two-ladder configuration while keeping the absorption negligible. Our simulations demonstrate the feasibility of achieving large cross-phase modulation in the system with small absorption.

Because slow light experiments can be performed under Autler–Townes or EIT conditions [29–31], which are often confused, it should be stressed that in this paper we discuss the case of narrow band operations, for which EIT is the most suitable [32].

Furthermore, we present an overview of the impact of parameters (excitonic states, control field intensities, temperature, or sample size) and provide a realistic example of a system to facilitate an experimental demonstration of XPM for RE in Cu_2O. The proposed scheme could possibly be used to implement photon–photon quantum gates, demonstrating the potential of Rydberg excitons media as a platform for quantum communications and quantum networking.

Our paper is organized as follows. In Section 2 we outline the theory of cross-phase modulation in an inverted Y system. Then, in Section 3 the results of calculations for a chosen excitonic state combination are presented and the impacts of various conditions on the system are discussed.

2. Theory

In the following, we consider a Cu_2O crystal as a medium with Rydberg excitons, where the XPM under conditions imposed by EIT in the linear regime can be realized. We use the so-called inverted Y configuration, which consists of two subsystems of ladder configuration (Figure 1). The whole system is composed of four levels a_1, a_2, b, and c. As in our previous works on Rydberg excitons [28], we focus our attention on the yellow excitonic series. We chose the valence band as the b state. As a practical example, the n_1P and n_2S excitonic states and dipole-allowed transitions WEre chosen. The description of the optically allowed transitions in this system is as follows: the ground state b, which is identified with the valence band, is coupled by two weak probe and signal beams of Rabi frequencies Ω_1 and Ω_3 with states a_1 and a_2. These two states are the sublevels of a state a obtained by applying a constant external magnetic field producing Zeeman splitting of the P-exciton levels [33], which in our case applies only to the a (n_1P) excitonic state. Note that these two weak beams are slightly detuned from the $a - b$ resonance. The two empty upper states a and c are coupled by the control field of Rabi frequency Ω_2. If one of the weak (signal or probe) fields is missing, the system reduces to a standard ladder-type three-level EIT configuration (in the linear case), driven by the control field, which has been considered in our previous paper [28]. With the three fields shown in Figure 1, our scheme acts as a double EIT system with two independent probe and signal fields propagating in the two transparency windows sharing the common control field. In such a situation, we deal with the case of multi-channel propagation under EIT conditions, so two different weak light pulses centered at two independent transparency frequencies travel with slow group velocities through the RE medium.

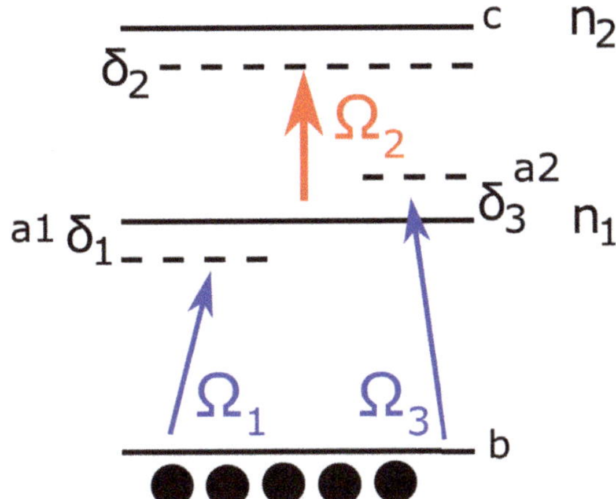

Figure 1. Schematic of the considered inverted Y configuration.

In order to study the propagation and interaction of the signal and probe fields inside the medium, the set Bloch equations of the form similar to that considered in our previous paper [28] (in the stationary case for the linear EIT regime) is accompanied by the Maxwell propagation equations for the Rabi frequency Ω_1 and Ω_3 of the probe and signal pulses, which in the slowly varying envelope approximation read

$$\left(\frac{\partial}{\partial t} + c\frac{\partial}{\partial z}\right)\Omega_{1,3}(z,t) = -i\kappa_{1,3}^2\sigma_{ba_{1,2}},\tag{1}$$

where $\kappa_{1,3}^2 = \frac{N|d_{a_{1,2}b}|^2\omega_{1,3}}{2\hbar\epsilon_0}$. N is the density of excitons and $d_{a_{1,3}b}$ are the transition dipole matrix elements of specific transitions, $\omega_{1,3}$ are the probe and signal frequencies and ϵ_0 is the vacuum dielectric permittivity. $\sigma_{ij}(z,t)$, $i,j \in \{a_1, a_2, b, c\}$ denotes the density matrix for an exciton at position z and time t.

$\Omega_{1,3}(z,t) = \frac{d_{a_{1,3}b}\varepsilon_{1,3}(z,t)}{\hbar}$ and $\Omega_2(z,t) = \frac{d_{ac}\varepsilon_2(z,t)}{\hbar}$ are the Rabi frequencies of the probe, signal and control fields corresponding to the particular couplings. The key of cross-Kerr nonlinearity lies in the fact that the phase of one light field is modified by an amount determined by the intensity of another optical field. The necessary conditions to achieve a significant cross-phase modulation (over π) are the small absorption and a steep dispersion, which are accomplished due to the EIT. A considerable reduction of the group velocities for both pulses traveling inside the medium allows these two optical fields to mutually interact in a common transparency window for a sufficiently long time. Rebic et al. have shown that by slightly departing from exact resonance conditions, one can obtain a group velocity matching and strong cross-Kerr modulation [34], which facilitates the phase gate operation in this system.

To theoretically describe the setup in the inverted Y configuration, we used the standard method to derive the formula for the susceptibilities, solving the set of stationary Bloch equations in the limit of low probe and signal intensities [35]. The derivation of susceptibility proceeds in a similar way as it was shown in [28]. We expand the system considered in [28] with an additional energy level and after solving the set of Bloch equations in the stationary regime (neglecting nonlinear potential), we obtain the susceptibilities for both probe and signal fields Ω_1 and Ω_3 in the form [36]:

$$\chi_1(\delta_1,\delta_3,\Omega_1,\Omega_3) = \frac{N|d_{ba_1}|^2}{\hbar\varepsilon_0}\frac{\sigma_{a_1 b}}{\Omega_1},$$

$$\chi_3(\delta_1,\delta_3,\Omega_1,\Omega_3) = \frac{N|d_{ba_2}|^2}{\hbar\varepsilon_0}\frac{\sigma_{a_2 b}}{\Omega_3}. \tag{2}$$

Because the probe and signal fields are weaker than the control one $|\Omega_1|^2, |\Omega_3|^2 << |\Omega_2|^2$, the above expressions for susceptibilities can be expanded into Taylor series

$$\chi_1 \approx \chi_1^{(1)} + \chi_{11}^{(3)}|\Omega_1|^2 + \chi_{13}^{(3)}|\Omega_3|^2,$$

$$\chi_3 \approx \chi_3^{(1)} + \chi_{33}^{(3)}|\Omega_3|^2 + \chi_{31}^{(3)}|\Omega_1|^2, \tag{3}$$

where $\chi_1^{(1)}$ and $\chi_3^{(1)}$ are the linear part of electric susceptibilities, $\chi_{11}^{(3)}$ and $\chi_{33}^{(3)}$ describe self-Kerr phase modulation (i.e., when an optical field modifies its own phase), and $\chi_{13}^{(3)}$, $\chi_{31}^{(3)}$ are responsible for cross-phase modulation. The various detunings and relaxation rates present in the system can be grouped in the following notation:

$$
\begin{aligned}
\Delta_{a_1 b} &= -\delta_1 + i\Gamma_{a_1 b},\\
\Delta_{cb} &= -\delta_2 + i\gamma_{cb},\\
\Delta_{a_2 b} &= -\delta_3 + i\Gamma_{a_1 b},\\
\Delta_{a_1 c} &= -\delta_1 - \delta_2 + i\Gamma_{a_2 b},\\
\Delta_{a_1 a_2} &= -\delta_1 - \delta_3,\\
\Delta_{ca_2} &= -\delta_3 - \delta_2 + i\Gamma_{a_2 b}.
\end{aligned}
\tag{4}
$$

where the parameters Γ_{ij}, $i \neq j$ describe the damping of exciton states and are determined by temperature-dependent homogeneous broadening due to phonons and broadening due to structural imperfections and eventual impurities. The relaxation damping rates for the coherence are denoted by $\gamma_{ij} \approx \Gamma_{ij}/2$, $i \neq j$ [37]. To simplify the expressions describing the susceptibility, we define the following functions of the probe field Ω_1:

$$
\begin{aligned}
O_1 &= \Omega_1 + \Delta_{a_1 b},\\
O_2 &= \Omega_1 + \Delta_{a_1 c},\\
O_3 &= \Omega_1 + \Delta_{a_2 b}^{*},\\
O_4 &= \Omega_1 + \Delta_{ca_2}.
\end{aligned}
\tag{5}
$$

After some calculations, we arrive at the linear, self-Kerr, and cross-Kerr parts of susceptibility (Equation (3)) in the following forms:

$$
\begin{aligned}
\chi_1^{(1)} &= \frac{\chi_0}{O_1 - \frac{\Omega_2^2}{O_2}},\\[2mm]
\chi_{11}^{(3)} &= \frac{0.5\chi_0\Omega_1^2}{\Delta_{ca_1}^{*}}\frac{\chi_0}{O_1 - \frac{\Omega_2^2}{O_2}},\\[2mm]
\chi_{13}^{(3)} &= \chi_0\left[\frac{\Delta_{a_2 b}^{*}}{\Delta_{cb}^{*}}\frac{1}{O_1 - \frac{\Omega_2^2}{O_2}}\left(\frac{0.5}{O_1 - \frac{\Omega_2^2}{O_2}} + \frac{0.5}{O_3 - \frac{\Omega_2^2}{O_4}}\right) - \frac{0.5}{\Delta_{ca_2}^{*}}\frac{\Omega_3^2}{(O_1 - \frac{\Omega_2^2}{O_2})(O_3 - \frac{\Omega_2^2}{O_4})}\right],
\end{aligned}
\tag{6}
$$

where χ_0 is a constant given by [27]

$$\chi_0 = \frac{N|d_{ba_i}|^2}{\hbar\epsilon_0}, \quad (i = 1, 2). \tag{7}$$

In analogy to the procedure presented above, with the following functions of the signal Rabi frequency Ω_3

$$\begin{aligned}
O_1' &= \Omega_3 + \Delta_{a_2b}, \\
O_2' &= \Omega_3 + \Delta_{ca_2}, \\
O_3' &= \Omega_3 + \Delta_{a_1b}^*, \\
O_4' &= \Omega_3 + \Delta_{ca_1},
\end{aligned} \tag{8}$$

one arrives at the expressions for the three parts of electric susceptibility for the signal field Ω_3

$$\begin{aligned}
\chi_3^{(1)} &= \frac{\chi_0}{O_1' - \frac{\Omega_2^2}{O_2'}}, \\
\chi_{33}^{(3)} &= \frac{0.5\chi_0\Omega_3^2}{\Delta_{ca_2}^*} \frac{\chi_0}{O_1' - \frac{\Omega_2^2}{O_2'}}, \\
\chi_{31}^{(3)} &= \chi_0 \left[\frac{\Delta_{a_1b}^*}{\Delta_{cb}^*} \frac{1}{O_1' - \frac{\Omega_2^2}{O_2'}} \left(\frac{0.5}{O_1' - \frac{\Omega_2^2}{O_2'}} + \frac{0.5}{O_3' - \frac{\Omega_2^2}{O_4'}} \right) - \frac{0.5}{\Delta_{ca_2}^*} \frac{\Omega_1^2}{(O_1' - \frac{\Omega_2^2}{O_2'})(O_3' - \frac{\Omega_2^2}{O_4'})} \right].
\end{aligned} \tag{9}$$

Although formulas for susceptibilities in Equations (6) and (9) have a complex form, their dependence on detunings is visible. The resonance or equal detunings give rise to similar dispersive properties for both 1 and 3 fields, while the nonlinear susceptibility vanishes and the XPM will not occur. Disturbing the EIT conditions by choosing different, but sufficiently small (to still remain in the common transparency window and preserving small absorption) detunings enables one to obtain a nonlinear contribution to susceptibility. The refraction indices $n_{1,3} = \sqrt{\epsilon_b + \chi_{1,3}}$, where $\epsilon_b = 7.5$ is the bulk permittivity of Cu_2O, together with the definition of the group velocity

$$v_g = \frac{c}{n_g} = \frac{c}{n + \omega\frac{dn}{d\omega}}, \tag{10}$$

where n_g is the group index, enables one to obtain the expressions for the group velocities of the propagating pulses

$$\begin{aligned}
v_g^{(1)} &= \frac{A}{\omega_1|d_{a_1b}|^2}(\Omega_2^2 + \Omega_3^2) \tag{11} \\
v_g^{(3)} &= \frac{A}{\omega_3|d_{a_2b}|^2}(\Omega_2^2 + \Omega_1^2),
\end{aligned}$$

where $A = \frac{4\pi c\epsilon_0}{N}$. From the above equations it follows that, because the orders of dipole moments are almost equal, changing the intensity of the control field and the probe or signal fields, it is possible to match in such a way their group velocities and therefore they can interact mutually in transparent medium for a sufficiently long time. The propagation equations for fields Ω_1 and Ω_3 have the following form:

$$\left(i\frac{\partial}{\partial z} + \frac{\delta_1}{c} - \frac{\Delta\omega_1}{c}\right)\Omega_1 = -\frac{\omega_1}{2c}\chi_1\Omega_1,$$

$$\left(i\frac{\partial}{\partial z} + \frac{\delta_3}{c} - \frac{\Delta\omega_3}{c}\right)\Omega_3 = -\frac{\omega_3}{2c}\chi_3\Omega_3,$$

where $\Delta\omega_i = \omega_2 - \omega_i$. Even inside the transparency window, a realistic medium is characterized by non-zero absorption coefficients

$$\alpha_{1,3} = \frac{\omega_{1,3}}{c} Im\sqrt{1 + \chi_{1,3}}. \tag{12}$$

Assuming that both pulses propagate in the z-direction through the sample of length L, their amplitudes are constant ($\Omega_{01,3} = \Omega_{1,3}(z = 0, \delta_{1,3}) = const$). The transmissions coefficients for probe and signal fields are defined by the following formulae:

$$T_1(\delta_1) = \frac{\Omega_1(L, \delta_1)}{\Omega_{10}},$$

$$T_3(\delta_3) = \frac{\Omega_3(L, \delta_3)}{\Omega_{30}}. \tag{13}$$

The phase difference is equal to the difference of optical path lengths

$$\phi_1(\omega) = \frac{(\omega_1 n_1 - \omega_3 n_3)L}{c}. \tag{14}$$

3. Numerical Results

As an example of an excitonic system where XPM can be realized, we used Cu_2O crystal of thickness $L = 200$ μm. The probe field coupled the ground state b (Figure 1) and sublevels of $n_1 = 2$, obtained by Zeeman split of the excitonic state in magnetic field. As a result, we obtained two levels shifted by $\delta_1 = -\delta_3 = 10$ GHz. The control field coupled the n_1 state with the empty upper state $n_2 = 10$. The exciton density was $N = 5.4 \times 10^{15}$ cm^{-3}, which was limited by the Rydberg blockade effect caused by the populated lower state n_1 [27]. In the case of $n_1 = 2$, the upper density limit was 2.6×10^{16} cm^{-3} [21]. The Rabi frequency of the control field was $\Omega_2 = 600$ GHz, which is comparable to the dissipation rate of the lower state $\gamma_{ab} = 2140$ GHz [38] for the temperature, $T = 10$ K. The calculated susceptibility is shown in the Figure 2. One can see that the real parts of both susceptibilities exhibited steep, normal dispersions, while the imaginary parts featured transparency windows in the form of dips. The transparency windows of both probe and signal fields overlapped, providing a common spectral region of small absorption, where the pulses can propagate. Due to the presence of detunings δ_1, δ_2, there was a noticeable frequency shift between both windows and dispersions, which resulted in a phase difference between propagating signals given by Equation (14). Figure 3a shows the group velocity index n_g of the signal field Ω_1 inside the transparency window as a function of control field Ω_2. One can see that there is an optimum value $\Omega_2/\gamma_{ab} \approx 0.15$, for which the slowdown was the strongest. For this strength of the control field, the transparency window was fully formed, but still narrow enough to provide a steep normal dispersion. The obtained slowdown was on the order of 10^4, which means that for the given sample thickness $L = 200$ μm, the propagation time through the crystal was $\tau \sim 7$ ns. This corresponds to the pulse spectral width $\Delta\omega \sim 15$ MHz, which is well below the width of the transparency window. Figure 3b shows the calculated cross-phase modulation. The maximum value of $\varphi_1 \sim 4.4$ rad was obtained in a wide range of control field strengths, centered around $\Omega_2 \approx 500$ GHz. As pointed out by Feizpour [39], the phase modulation scales as $1/\Delta_{EIT}$, where Δ_{EIT} is the spectral width of the window, provided that the window is wide enough and is limited by the strength of the control field Ω_2 and dissipation rate γ_{ab}. In principle, one can use higher excitonic states with smaller γ_{ab} to obtain a much narrower transparency window. However, in our calculations this benefit was offset with significantly smaller exciton density. This in turn resulted in a smaller value of susceptibility which produced a smaller phase shift. However, a large real part of susceptibility is accompanied by a significant imaginary part which results in absorption. Due to the particularly large dissipation constants as compared to atomic EIT system, one can observe in Figure 2 that even inside the transparency window $\mathrm{Im}\,\chi \sim 10^{-3}$, which resulted in absorption on the order of $\alpha \sim 20\ \mathrm{cm}^{-1}$. According to Equation (13), this corresponds to about 70% transmission through a $L = 200\ \mu\mathrm{m}$ sample. The absorption coefficient is consistent with experimental results by Malerba et al. [40], where the measured values outside the resonance peaks were in the range of 10^1–$10^2\ \mathrm{cm}^{-1}$, depending on sample thickness. To sum up, the imaginary part of χ inside the transparency window provided a contribution to absorption that was on the same order as the intrinsic, bulk absorption due to the defects [40] and was sufficiently low to ensure considerable transmission. Since the phase shift was directly proportional to L, there was an interplay between XPS and signal transmission.

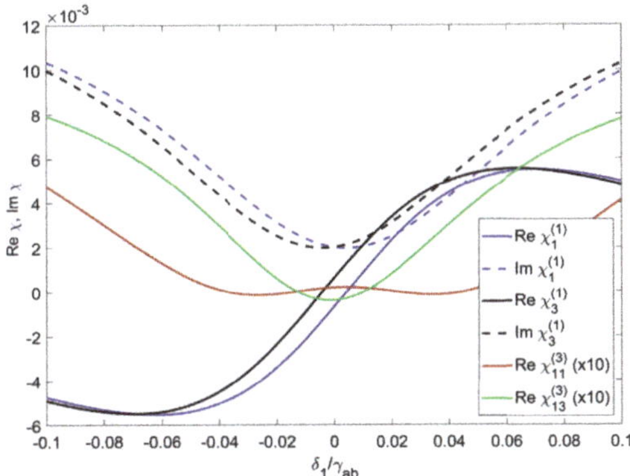

Figure 2. Linear and nonlinear parts of signal field susceptibilities (real and imaginary part) $\chi_1^{(1)}$, $\chi_3^{(1)}$, $\chi_{11}^{(3)}$, $\chi_{13}^{(3)}$ for both probe and signal fields. The Rabi frequencies are $\Omega_2 = 600\ \mathrm{GHz}$, $\Omega_1 = \Omega_3 = 60\ \mathrm{GHz}$, exciton density is $N = 5.4 \times 10^{15}\ \mathrm{cm}^{-3}$.

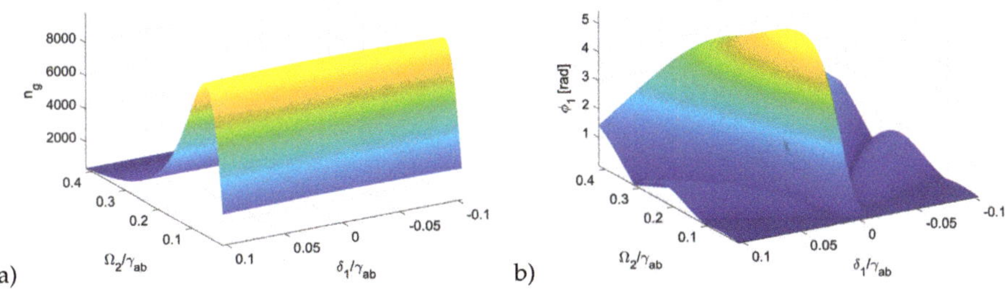

Figure 3. (**a**) Group velocity index and (**b**) cross-phase modulation as a function of detuning δ_1 and control field Rabi frequency Ω_2, for the same parameters as in Figure 2.

Finally, we investigated how the XPS scaled with temperature by applying the excitonic line-broadening model described in [27] to the system described above. Figure 4a depicts the maximum slowdown as a function of temperature. For the chosen transparency window width, the slowdown was largely unaffected by broadening up to $T \sim 40\ \mathrm{K}$. Likewise, the cross-phase modulation shown in Figure 4b exhibits identical behavior. This result is consistent with the findings presented in [39], where a similar dependence of XPS on the dephasing rate is shown. Notably, the slowdown and XPS remained significant even at $T = 100$ K. This is possible mainly due to the choice of a low-lying state $n_1 = 2$. As mentioned before, for upper states, the optimal results are obtained with much narrower transparency windows, which are more disturbed by line broadening, which is also much greater for these higher Rydberg states.

We emphasize that all presented numerical results are based on a realistic and experimentally verified parameters for RE in Cu_2O. We used the usual theoretical approach to derive the formula for the susceptibilities by solving the set of stationary Bloch equations in the low range of probe and signal intensities [36]. The calculated values of cross-phase modulation represent a similar dependence on the control field intensity as those measured recently by Sinclair et al. [41] in a cold Rubidium gas.

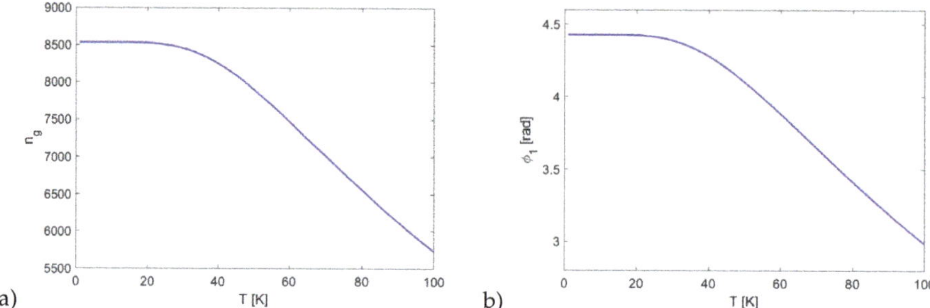

Figure 4. (**a**) Group velocity index and (**b**) cross-phase modulation as a function of temperature.

4. Conclusions

In this paper, we studied the nonlinear response of Rydberg excitons in Cu_2O sample in an inverted Y configuration, where the incident probe and signal fields interact in EIT conditions. By expanding the ladder system presented in [28] with additional signal field and adjusting the parameters to enter the nonlinear regime, we derived expressions for the third-order susceptibility and suggested the optimal set of parameters for which the remarkable nonlinearities in Cu_2O with RE might be experimentally realized. Rydberg excitons in Cu_2O have now reached a stage at which the coherent quantum effects and controlled quantum manipulations could be realized. With Rydberg atoms, it has been possible to obtain a large optical nonlinearity at the single photon level and perform many sophisticated quantum optics experiments such as optical Kerr effect or correlated states [11]. It is expected that the medium of Rydberg excitons is also a fertile area [22,23,32,35,41].

We have demonstrated that it is possible in principle to achieve a phase difference of over π in a 200 µm sample, at temperatures approaching 100 K and exciton densities an order of magnitude below the limit imposed by Rydberg blockade. Since their discovery in 2014, the Rydberg excitons in Cu_2O have been investigated mostly from the spectroscopic point of view while only a few experiments have focused on their quantum optical applications [22,23], which have confirmed their usefulness in quantum information processing. We hope that our investigations will help in the use of REs as intermediaries in photon–matter coupling in the field of modern quantum processing in solids.

Author Contributions: Conceptualization, D.Z. and S.Z.-R.; Investigation, S.Z.-R. and D.Z.; Methodology, S.Z.-R. and D.Z.; Supervision, S.Z.-R.; Writing—original draft, D.Z. and S.Z.-R. All authors have read and agreed to the published version of the manuscript.

Funding: Support from the National Science Centre, Poland (project OPUS 2017/25/B/ST3/00817) is greatly acknowledged.

Conflicts of Interest: The authors declare no conflict of interest.

References

1. Fleischhauer, M.; Immamoglu, A.; Marangos, J.P. Electromagnetically induced transparency: Optics in coherent media. *Rev. Mod. Phys.* **2005**, *77*, 633. [CrossRef]
2. Distante, E.; Padrón-Brito, A.; Cristiani, M.; Paredes-Barato, D.; de Riedmatten, H. Storage Enhanced Nonlinearities in a Cold Atomic Rydberg Ensemble. *Phys. Rev. Lett.* **2016**, *117*, 113001. [CrossRef]
3. Schmidt, H.; Imamoglu, M. Giant Kerr nonlinearities obtained by electromagnetically induced transparency. *Opt. Lett.* **1996**, *21*, 1936. [CrossRef] [PubMed]
4. Shiau, B.; Wu, M.; Lin, C.; Chen, Y. Low-Light-Level Cross-Phase Modulation with Double Slow Light Pulses. *Phys. Rev. Lett.* **2011**, *106*, 193006. [CrossRef] [PubMed]
5. Bai, Z.; Huang, G. Enhanced third-order and fifth-order Kerr nonlinearities in a cold atomic system via Rydberg-Rydberg interaction. *Opt. Express* **2016**, *24*, 4442. [CrossRef] [PubMed]
6. Vitali, D.; Fortunato, M.; Tombesi, P. Complete Quantum Teleportation with a Kerr Nonlinearity. *Phys. Rev. Lett.* **2000**, *85*, 445. [CrossRef] [PubMed]
7. Munro, W.; Nemoto, K.; Spiller, T. Weak nonlinearities: A new route to optical quantum computation. *New J. Phys.* **2005**, *7*, 137. [CrossRef]
8. Hosseini, M.; Rebić, S.; Sparkes, B.; Twamley, J.; Buchler, B.; Lam, K.P. Memory-enhanced noiseless cross-phase modulation. *Light Sci. Appl.* **2012**, *1*, e40. [CrossRef]
9. Volz, J.; Scheucher, M.; Junge, C.; Rauschenbeutel, A. Nonlinear π phase shift for single fibre-guided photons interacting with a single resonator-enhanced atom. *Nat. Photonics* **2014**, *8*, 965. [CrossRef]
10. Tiarks, D.; Schmidt, S.; Rempe, G.; Dürr, S. Optical π phase shift created with a single-photon pulse. *Sci. Adv.* **2016**, *2*, e1600036. [CrossRef]
11. Firstenberg, O.; Adams, C.; Hofferberth, S. Nonlinear quantum optics mediated by Rydberg interactions. *J. Phys. B At. Mol. Opt. Phys.* **2016**, *49*, 152003. [CrossRef]
12. Johnsson, M.; Molmer, K. Storing quantum information in a solid using dark-state polaritons. *Phys. Rev. A* **2004**, *70*, 032320. [CrossRef]
13. Heinze, G.; Hubrich, C.; Halfmann, T. Stopped Light and Image Storage by Electromagnetically Induced Transparency up to the Regime of One Minute. *Phys. Rev. Lett.* **2013**, *111*, 033601. [CrossRef] [PubMed]
14. Schraft, D.; Hain, M.; Lorenz, N.; Halfmann, T. Stopped Light at High Storage Efficiency in a Pr3+:Y2SiO5 Crystal. *Phys. Rev. Lett.* **2016**, *116*, 073602. [CrossRef] [PubMed]
15. Fuchs, G.; Burkard, G.; Klimov, P.; Awschalom, D. A quantum memory intrinsic to single nitrogen vacancy centres in diamond. *Nat. Phys.* **2011**, *7*, 789. [CrossRef]
16. Ian, H.; Liu, Y.; Nori, F. Tunable electromagnetically induced transparency and absorption with dressed superconducting qubits. *Phys. Rev. A* **2010**, *81*, 063823. [CrossRef]
17. Sun, H.; Liu, Y.; Ian, H.; You, J.; Il'ichev, E.; Nori, F. Electromagnetically induced transparency and Autler-Townes splitting in superconducting flux quantum circuits. *Phys. Rev. A* **2014**, *89*, 063822. [CrossRef]
18. Wang, H.; Gu, X.; Liu, Y.; Miranowicz, A.; Nori, F. Optomechanical analog of two-color electromagnetically-induced transparency: Photon transmission through an optomechanical device with a two-level system. *Phys. Rev. A* **2014**, *90*, 023817. [CrossRef]
19. Jing, H.; Özdemir, S.; Geng, Z.; Zhang, J.; Lü, X.; Peng, B.; Yang, L.; Nori, F. Optomechanically-induced transparency in parity-time-symmetric microresonators. *Sci. Rep.* **2015**, *5*, 9663. [CrossRef]
20. Heckötter, J.; Freitag, M.; Fröhlich, D.; Aßmann, M.; Bayer, M.; Semina, M.; Glazov, M. Scaling laws of Rydberg excitons. *Phys. Rev. B* **2017**, *96*, 125142. [CrossRef]
21. Kazimierczuk, T.; Fröhlich, D.; Scheel, S.; Stolz, H.; Bayer, M. Giant Rydberg excitons in the copper oxide Cu_2O. *Nature* **2014**, *514*, 344. [CrossRef] [PubMed]
22. Grünwald, P.; Aßmann, M.; Heckötter, J.; Fröhlich, D.; Bayer, M.; Stolz, H.; Scheel, S. Signatures of Quantum Coherences in Rydberg Excitons. *Phys. Rev. Lett.* **2016**, *117*, 133003. [CrossRef] [PubMed]
23. Khazali, M.; Heshami, K.; Simon, C. Single-photon source based on Rydberg exciton blockade. *J. Phys. B* **2017**, *50*, 215301. [CrossRef]
24. Walther, V.; Johne, R.; Pohl, T. Giant optical nonlinearities from Rydberg excitons in semiconductor microcavities. *Nat. Commun.* **2018**, *9*, 1309. [CrossRef]
25. Walther, V.; Kruger, S.; Scheel, S.; Pohl, T. Interactions between Rydberg excitons in Cu_2O. *Phys. Rev. B* **2018**, *98*, 1165201. [CrossRef]

26. Ziemkiewicz, D.; Zielińska-Raczyńska, S. Proposal of tunable Rydberg exciton maser. *Opt. Lett.* **2018**, *43*, 3742–3745. [CrossRef]
27. Ziemkiewicz, D.; Zielińska-Raczyńska, S. Solid-state pulsed microwave emitter based on Rydberg excitons. *Opt. Express* **2019**, *27*, 16983. [CrossRef]
28. Ziemkiewicz, D. Electromagnetically Induced Transparency in media with Rydberg Excitons 1: Slow light. *Entropy* **2019**, *22*, 177. [CrossRef]
29. Anisimov, P.; Dowling, J.; Sanders, B. Objectively Discerning Autler-Townes Splitting from Electromagnetically Induced Transparency. *Phys. Rev. Lett.* **2011**, *107*, 163604. [CrossRef]
30. Peng, B.; Özdemir, S.; Chen, W.; Nori, F.; Yang, L. What is and what is not electromagnetically induced transparency in whispering-gallery microcavities. *Nat. Commun.* **2014**, *5*, 5082. [CrossRef]
31. Liu, Q.C.; Li, T.; Luo, X.; Zhao, H.; Xiong, W.; Zhang, Y.; Chen, Z.; Liu, J.; Chen, W.; Nori, F.; Tsai, J.; You, J. Method for identifying electromagnetically induced transparency in a tunable circuit quantum electrodynamics system. *Phys. Rev. A* **2016**, *93*, 053838. [CrossRef]
32. Rastogi, A.; Saglamyurek, E.; Hrushevskyi, T.; Hubele, S.; LeBlanc, L. Discerning quantum memories based on electromagnetically-induced-transparency and Autler-Townes-splitting protocols. *Phys Rev. A* **2019**, *100*, 012314. [CrossRef]
33. Zielińska-Raczyńska, S.; Ziemkiewicz, D.; Czajkowski, G. Magneto-optical properties of Rydberg excitons: Center-of-mass quantization approach. *Phys. Rev. B* **2017**, *95*, 075204. [CrossRef]
34. Rebić, S.; Vitali, D.; Ottaviani, C.; Tombesi, P.; Artoni, M.; Cataliotti, F.; Corbalán, R. Polarization phase gate with a tripod atomic system. *Phys. Rev. A* **2004**, *70*, 032317. [CrossRef]
35. Slowik, K.; Raczyński, A.; Zaremba, J.; Zielińska-Kaniasty, S.; Artoni, M.; La Rocca, G. Cross-phase modulation and population redistribution in a periodic tripod medium. *J. Mod. Opt.* **2011**, *58*, 978–987. [CrossRef]
36. Scully, M.; Zubairy, M. *Quantum Optics*; Cambridge University Press: Cambridge, UK, 1997.
37. Artoni, M.; La Rocca, G.; Bassani, F. Electromagnetic-induced transparency of Wannier-Mott excitons. *Europhys. Lett.* **2000**, *49*, 445. [CrossRef]
38. Zielińska-Raczyńska, S.; Ziemkiewicz, D.; Czajkowski, G. Electro-optical properties of Rydberg excitons. *Phys. Rev. B* **2016**, *94*, 045205. [CrossRef]
39. Feizpour, A.; Dmochowski, G.; Steinberg, A. Short-pulse cross-phase modulation in an electromagnetically-induced-transparency medium. *Phys. Rev. A* **2016**, *93*, 013834. [CrossRef]
40. Malerba, C.; Biccari, F.; Ricardo, C.; D'Incau, M.; Scardi, P.; Mittiga, A. Absorption coefficient of bulk and thin film Cu_2O. *Sol. Energy Mater. Sol. Cells* **2011**, *95*, 2848–2854. [CrossRef]
41. Sinclair, J.; Daniela Angulo, D.; Lupu-Gladstein, N.; Bonsma-Fisher, K.; Steinberg, A.M. Observation of a large, resonant, cross-Kerr nonlinearity in a cold Rydberg gas. *Phys. Rev. Res.* **2019**, *1*, 033193. [CrossRef]

 *entropy*

Article

Fractal Plasmons on Cantor Set Thin Film

David Ziemkiewicz *, Karol Karpiński and Sylwia Zielińska-Raczyńska

Institute of Mathematics and Physics, UTP University of Science and Technology, Al. Prof. S. Kaliskiego 7, 85-789 Bydgoszcz, Poland; karol.karpinski@utp.edu.pl (K.K.); sziel@utp.edu.pl (S.Z.-R.)
* Correspondence: david.ziemkiewicz@utp.edu.pl

Received: 5 November 2019; Accepted: 27 November 2019; Published: 29 November 2019

Abstract: The propagation of surface plasmon–polaritons is investigated in a metallic, fractal-like structure based on Cantor set. The dynamic of plasmonic modes generating on the Cantor structure is discussed in the context of the setup geometry. The numerically obtained reflection spectra are analyzed with the box-counting method to obtain their dimension, which is shown to be dependent on the geometry of the plasmonic structure. The entropy of the structure is also calculated and shown to be proportional to the dimension. Presented analysis allows for extracting information about fractal plasmonic structure from the reflectance spectrum. Predictions regarding the experimental observation of discussed effects are presented.

Keywords: surface plasmons; fractals; entropy

---

1. Introduction

Over the past decades, the study of electromagnetic waves propagating at the interface between a metal and dielectric has been of significant interest. In such a physical system the energy of the electromagnetic field is confined to sub-wavelength size supporting localized surface plasmon polariton (SPP) modes. Their unusual properties depend on the metal and geometry of the material structure, which in fact plays a crucial role in local electromagnetic enhancement and tuning the resonance positions. Such interdependences have been recently studied regarding the impact of the nonlocality of the surface structure on the dispersion relation of SPP [1]. A lot of work has been devoted to the examination of plasmonic properties of various metal surfaces with specific geometrical shapes, e.g., periodic V-grooves [2], slabs with periodic grating [3] or to investigate propagation of a confined field in periodically corrugated waveguides [4]. Carefully designed metasurfaces offer new opportunities for tunable unidirectional excitation of SPPs whose propagating direction depends on the helicity of incident light [5]. The recent advancements in the research of geometric metasurfaces and their applications in ultrathin optical devices have been presented in [6]. The idea of examining fractal geometries was a natural step forward in plasmonic studies. In recent years there has been continuous interest in the development of fractal metamaterials, which could be used in high-gain, compact, multiband antennas [7–9].

Significant research effort has been devoted to fractal space-time systems, starting from analysis of diffusion in fractal space [10,11]. Interesting results can be obtained even in simple wave propagation problems; Berry has proved that the diffraction patterns caused by fractal objects are fractals themselves [12], which has been since experimentally confirmed [13]. Cherny et al. [14] studied the small-angle scattering in generalized Cantor set fractals.

There is an inherent connection between entropy and fractal dimension [15]; recently, Chen [16] has used this connection for the study of urban systems, which are natural, self-organizing, fractal structure. While there are some studies of entropy of electromagnetic waves reflected from metallic structure [17], it seems that this paper is the first attempt to link the relevance of entropy of fractal systems where surface plasmons are excited.

It is important to note that realistic, fractal-like physical systems cannot contain infinitely small parts; if the construction is iterative, only a finite number of iterations is used. Examples of such structure are triadic-Cantor photonic crystals [18] and plasmonic superlattices [19] which can be analyzed using standard transfer matrix approach. Similarly, the finite number of Cantor set iterations in plasmonic structure allows one to approach the problem with classical optics and to use the standard form of Fourier transform.

These observations and studies inspire us to study plasmon propagation in fractal structures. Such systems have geometry which scales in a nontrivial way and is characterized by a non-integer dimension. One of the simplest examples of a fractal is the Cantor set [20]. The aim of our paper is to investigate the propagation of plasmons generated by an electromagnetic wave incident at the quasi-periodic Cantor-like metallic layer. In our analysis, the structure is generated in finite, but sufficient number of iterations so that the finest structures are much smaller than the wavelength. This is particularly important for plasmonic structure, as it is well known that SPPs can interact with sub-wavelength surface features. Therefore, one may expect a very rich system dynamics in the case of fractal metallic structures. By using the Cantor set as a geometrically simple example of fractal structure, we use the well-established principles of SPP excitation in optical dipole antennas [21] and thin layers [22] to predict the number and frequency of SPP modes excited in the system. Then, we demonstrate that the reflection spectrum of fractal plasmonic structure can be interpreted as a fractal and characterized by a non-integer dimension, providing a new, useful measure in spectroscopy of plasmonic systems. Our work may unlock a novel opportunity to elicit information about the surface structure from the reflection spectrum of SPPs and therefore establishes a link between optical properties and different fractal geometries. The results are general and applicable to a wide class of natural and fabricated systems characterized by a non-integer dimension. Although our considerations are purely theoretical, nevertheless the numerical simulations are based on the realistic model of the fractal-like structure of the silver film on the glass substrate; we believe that presented results can be an inspiration for the experimental investigations.

2. Theory

Our setup consists of a metal layer with indentations, deposited on a glass substrate (Figure 1a). The minimum and maximum layer thickness are d and $d + h$ respectively. The indentations are cut according to the iterative process of generalized Cantor set construction [20] (Figure 1); at every iteration, a middle part of the unmodified metal surface is cut, forming two thicker areas ("islands" with length a) and a middle, thinner area ("groove", with length b). The size of the removed part is given by a fraction $0 < f < 1$. Assuming that the total length of the structure is l, the length of grooves and islands obtained in the i-th iteration is respectively

$$
\begin{aligned}
a_i &= l\left[(1-f)/2\right]^i, \\
b_i &= lf\left[(1-f)/2\right]^{i-1}.
\end{aligned}
\tag{1}
$$

In our manuscript, we analyze the case in which the number of iterations is $i = 3$ and $f \in (0, 0.4)$, where $f = 1/3$ is the standard Cantor set. For $f > 0.4$ and $i > 3$, the size of the metallic parts quickly decreases, approaching the size of the unit cell in numerical simulation.

A TM-polarized incident wave (I on the Figure 1a) illuminates the surface from the glass side, at an angle $\theta = 45°$, in the Kretschmann configuration [23]. For some selected frequencies, the incident wave can excite multiple surface plasmon modes (marked by red and blue lines). One can expect that the strongest plasmonic resonances will have the form of standing waves, with multiples of half-wavelength matching the length of various horizontal surfaces b_i in the structure, e.g.,

$$
N\lambda/2 = b_i, \; N \in \mathbb{N}, \; i = 0\ldots n,
\tag{2}
$$

where i is the iteration number, so that the distances range from the total structure length $b_0 = l$ down to the smallest features b_n. Figure 2 shows the typical plasmonic modes obtained in a finite-difference time-domain (FDTD) simulation.

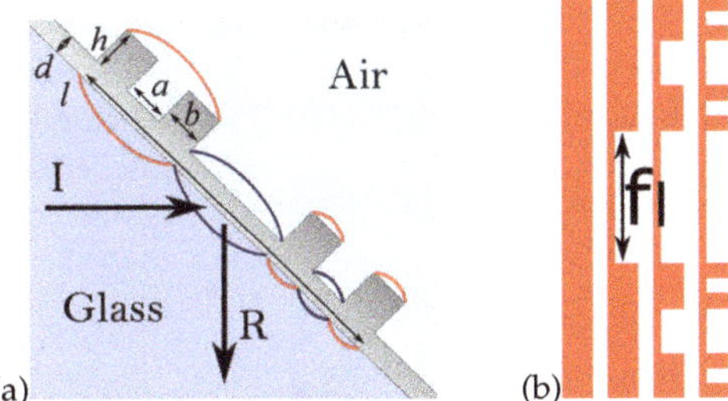

Figure 1. (**a**) Schematics of the considered plasmonic system, with propogation direction of incident (I) and reflected (R) waves. (**b**) First three iterations of Cantor set construction with structure length l and removed part fraction $f = 1/3$.

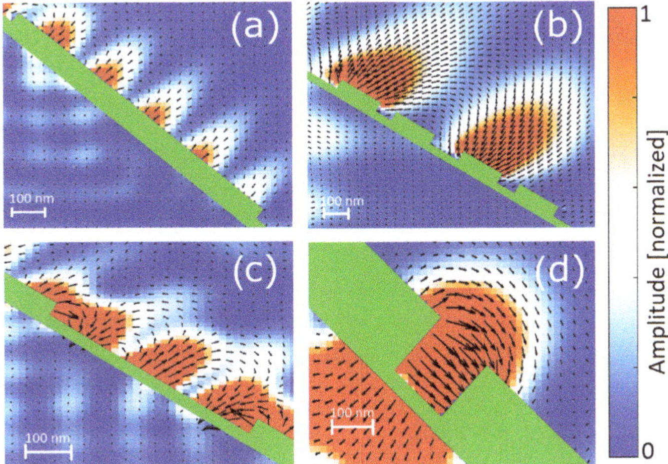

Figure 2. Various plasmonic modes forming on the Cantor structure. Color and arrows mark the electric field intensity E^2 and direction, respectively. The metallic structure is shown in green. (**a**) Island mode with $b = 2.5\lambda$. (**b**) Low-frequency island mode spanning whole first iteration structure, ignoring the second and third iteration grooves. (**c**) Groove mode consisting of standing wave with $a \approx 1.5\lambda$ and two additional intensity peaks on the island edges. (**d**) Narrow groove mode.

In Figure 2a one can see an example of "island" mode in the form of a standing wave, with wavelength being a fraction of the island length b. Typically for SPP, the electric field is perpendicular to the metal surface, drops exponentially with the distance from the metal and vanishes inside the metal. Interestingly, long-wavelength island modes can form over the grooves (Figure 2b). This means that in a Cantor structure created in i iterations, there will be SPP resonances corresponding to the structures obtained in iteration $i' \leq i$ and the whole reflection spectrum is a sum over all iterations. Figure 2c depicts a typical "groove" mode. A standing wave is formed inside the groove. In addition, there are two intensity peaks forming on the island edges enclosing the groove. Due to this phenomenon, the efficient wavelength of the model is slightly longer than the allowed space inside the groove could permit. This effect becomes more pronounced as the groove size a decreases (Figure 2d); in such a case, the plasmonic mode changes the field configuration, with the electric field direction parallel to the metal-glass interface. Moreover, the wavelength is no longer dependent on the groove length and instead becomes proportional to the groove depth of h. One can also see that this type of mode is relatively weak—the electric field amplitude is comparable to the free-propagating wave on the other side of the metal surface. In contrast, the other depicted SPP modes are much stronger than the incident and reflected field. This is caused by the fact that for those modes, the electric field on the vertical walls (edges) of the structure is negligible, reducing power losses due to induced currents flowing along these walls and thus creating strong, standing wave patterns. The excitation of plasmons takes energy from the incident wave, reducing the amplitude of the reflected wave R.

For quantitative results, we recall the model presented in [24], where the incident field excites SPPs on a silver layer. The permittivity of silver is calculated with the Drude model

$$\epsilon_m(\omega) = 1 - \frac{\omega_p^2}{\omega^2 + i\gamma\omega}, \tag{3}$$

with the plasma frequency $\omega_p = 1930.5$ THz and damping ratio $\gamma = 31.35$ THz obtained from fitting to the value $\epsilon_m(\lambda = 589\ nm) = -13.3 + 0.883i$ [25]. The metal film is deposited on a glass substrate with $\epsilon_g = 2.25$. To obtain the frequency of SPP modes, one has to consider the boundary conditions on the glass-metal and metal-air interfaces [22,26].

$$\frac{\kappa_m \tanh\left(\kappa_m \frac{d'}{2}\right)}{\epsilon_m} = \frac{-\kappa_a}{\epsilon_a} = \frac{-\kappa_g}{\epsilon_g},$$

$$\frac{\kappa_m \coth\left(\kappa_m \frac{d'}{2}\right)}{\epsilon_m} = \frac{-\kappa_a}{\epsilon_a} = \frac{-\kappa_g}{\epsilon_g}, \tag{4}$$

where $\kappa_j = \sqrt{k^2 - \epsilon_j \frac{\omega^2}{c^2}}$ denotes the component of the wave vector k parallel to the surface of the j-th medium with $j \in \{a, m, g\}$ corresponding respectively to the air, metal and glass; ϵ_j are the relative electric permittivities of these substances and d' is the layer thickness ($d' = d$ and $d' = d + h$ for groove and island modes, respectively). The two above equations describe the so-called short-range and long-range SPPs. In this paper, we consider only the stronger, long-range modes. Furthermore, in the case of the Cantor set structure shown on the Figure 1a, there are two different types of modes corresponding to the layer thickness d and $d + h$ respectively. By solving the Equation (4) numerically, we obtain a dispersion relation in a form $\omega(\lambda)$. This allows us to calculate the frequency of the modes with wavelengths given by (2), corresponding to the various distances in the structure given by Equation (1). Due to this close correspondence between the structure and the surface plasmons, one can expect that the number and frequency of the excited modes will reflect the recursive pattern of the metallic layer; overall, a feature-rich surface will result in feature-rich reflection spectrum due to the multiple SPP resonances. Dettmann [27] has used a Fourier transform to describe the Cantor set charge distribution and its accompanying electrostatic potential, which has been shown to be a dimension-dependent power law. Thus, the non-integer fractal dimension of the structure is a quantity that might be preserved to some degree in the reflection spectrum. However, the exact analytical relation between structure and spectrum dimension is nontrivial; every metallic surface supports multiple plasmonic modes, generating many overlapping spectral lines of various strength and width. The proposed system is one of the simplest models of a structure where the dimension can be precisely controlled with parameter f and the reflection spectrum is a result of well-known resonances given by (2), providing a convenient model and a basis for our further dimension analysis in systems with more complicated geometry.

3. Numerical Results

To calculate the reflection and scattering spectra, we have used the FDTD method, with medium parameters described in [24,26]. The simulation domain is a two-dimensional cross-section of the structure, as shown in Figure 1a. The whole domain is divided by a rectangular grid with a single cell size Δx. At every grid point, the electric and magnetic field distributions $\vec{E}(x,y,t)$, $\vec{H}(x,y,t)$ are calculated from their previous values $\vec{E}(x,y,t-\Delta t)$, $\vec{H}(x,y,t-\Delta t)$ with evolution equations derived directly from Maxwell's equations. In the chosen, two-dimensional system, one has three non-zero field components $\vec{E} = [E_x, E_y, 0]$ and $\vec{H} = [0, 0, H_z]$. The computational domain is a square grid, 1000×1000 unit cells. For this geometry, the most efficient use of available space is to place the metallic layer at the diagonal and use 45-degree incidence angle. In such a system, the incident field propagates only along x axis and has a single non-zero electric field component E_y whereas the reflected beam electric field is E_x. This allows for easy decoupling of incident and reflected fields for further analysis. The domain is terminated with absorbing layers having a reflection coefficient smaller than 10^{-4}. The structure is finite and contained within the domain in (x, y) plane. Since the simulation is two-dimensional, the system is assumed to be semi-infinite (much larger than the computational domain) in the z direction. The glass is assumed to have a constant susceptibility $\epsilon_g = 2.25$ while the metal is described by the Drude model (3). We have used ADE (Axillary Differential Equations) approach [28], where we compute medium polarization by solving a second-order PDE in the form

$$\ddot{P} + \gamma \dot{P} = \frac{\omega_p^2}{\epsilon_\infty} E, \tag{5}$$

with the fitted medium parameters: plasma frequency ω_p, damping constant γ and high frequency susceptibility limit ϵ_∞. The full set of equations solved in our FDTD approach is as follows

$$
\begin{aligned}
\frac{\partial E_y(x,y,t)}{\partial x} - \frac{\partial E_x(x,y,t)}{\partial y} &= -\mu_0 \frac{\partial H_z(x,y,t)}{\partial t}, \\
\frac{\partial H_z(x,y,t)}{\partial x} &= -j_y(x,y,t) - \epsilon_0 \frac{\partial E_y(x,y,t)}{\partial t} - \frac{\partial P_y(x,y,t)}{\partial t}, \\
\frac{\partial H_z(x,y,t)}{\partial y} &= j_x(x,y,t) + \epsilon_0 \frac{\partial E_x(x,y,t)}{\partial t} + \frac{\partial P_x(x,y,t)}{\partial t},
\end{aligned}
\tag{6}
$$

where P_x, P_y are components of polarization vector calculated from (5), j is the current density, ϵ_0, μ_0 are the vacuum permittivity and permeability. The above equations are rearranged to obtain time derivatives of the E_x, E_y, H_z fields, which are then used to calculate the field evolution with some constant time step Δt.

As mentioned before, due to the finite spatial resolution of the simulation Δx limiting the size of the smallest features, the Cantor set structure is generated in 3 iterations.

The Figure 3 shows the reflection spectrum in a system where $d = 0$ and $h = 45$ nm, e.g., the indentations are cut through the whole metal layer, forming islands of metal on the glass substrate. The range of values of f describes structures varying from continuous metal layer ($f = 0$) to small islands covering 21.6% of glass surface ($f = 0.4$). One can see multiple minima of reflection (shown in white and blue color), which correspond to the predicted plasmonic modes marked by dashed lines. These frequencies are calculated as follows: for any given length of metallic structure b_i, there are multiple matching standing-wave modes given by Equation (2). By using the dispersion relation (4), we calculate the mode frequencies $\omega_n(\lambda_n)$. The calculated modes are identified in Figures 3 and 4 by their wavelength shown on the top. In Figure 3 one can see that the frequencies are increasing with a fraction f due to the fact that the metal islands and corresponding plasmon wavelengths become smaller. As mentioned before, the best fit is obtained when taking into account the sizes of structures in multiple Cantor iterations. Additionally, the spectrum contains a wide minimum (thick, black line, marked l) which is a generic plasmonic mode of an uncut metal layer [24]. The large area of low reflection coefficient for $\omega > 500$ THz and $f > 0.1$ is a result of many overlapping plasmonic lines in this region; one can see that the local minima of reflection correspond to the crossing points of these lines. Overall, the reflection coefficient is proportional to the surface of metal and thus decreases with f. In the limit of $f = 0$, only generic, continuous layer mode and the lowest order island modes b_1, corresponding to the whole structure size l, are present. The close correspondence between reflection minima and predicted SPP resonances is more apparent in the cross-section shown in Figure 3 bottom panel. The two strongest resonances are the generic line overlapping with $\lambda = b_2/3$ line and the combination of b_2 and $b_1/2$ lines.

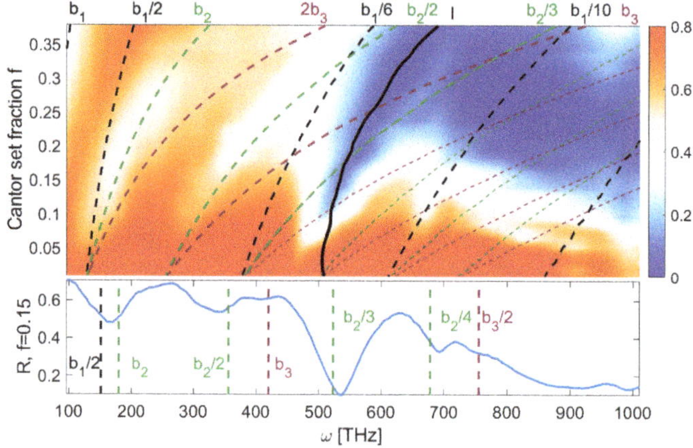

Figure 3. Reflection coefficient as a function of frequency and the Cantor set fraction f, for $d = 0$ and $h = 45$ nm. Black, green and magenta lines mark the plasmonic modes corresponding to first, second and third iteration structures.

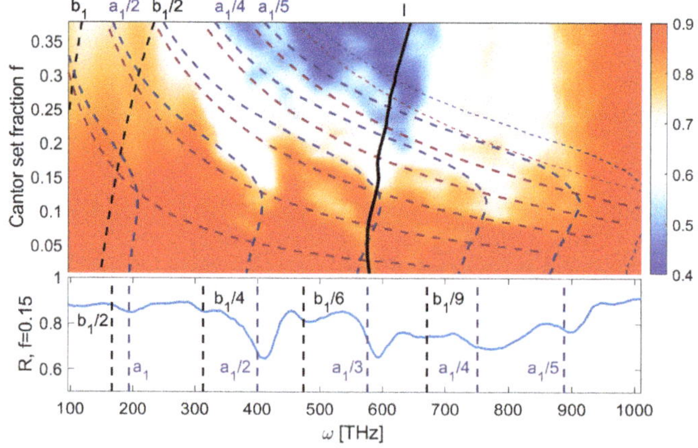

Figure 4. Reflection coefficient as a function of frequency and the Cantor set fraction f, for $d = 45$ nm and $h = 90$ nm. Black lines mark the island modes. Blue and magenta lines are the groove modes with and without low f correction, respectively.

The next analyzed system consists of deep grooves with $h = 90$ nm cut in a thick metal layer, resulting in minimum thickness $d = 45$ nm. In this case, one can expect not only the "island" modes, but also "groove" modes forming inside the indentations. This is shown in Figure 4. The reflection spectrum depends mostly on the groove modes, which are the strongest plasmonic resonances due to the fact that the metal layer is thinnest inside the indents. One could expect that in the limit $f \to 0$, the indents and corresponding wavelengths λ_i become negligibly small, and $\omega_i \to \infty$. Such modes are marked by magenta lines. However, as the grooves become narrower, the plasmonic modes start to form on their vertical sides, with constant wavelength $N\lambda/2 = h$. We have taken this into account by introducing effective groove size l which approaches h as $f \rightarrow 0$. This results in the characteristic shape of the blue lines seen in Figure 4, which are not divergent for $f \rightarrow 0$. The generic mode, again marked by thick, black line, is relatively weak due to the large effective thickness of the metal layer, especially in the limit of $f \rightarrow 0$. On the cross-section at the bottom of the figure, one can notice that some reflection minima are blue-shifted as compared to the predicted frequencies; as shown by Novotny [21], there is a correlation between a shape of emitting elements and wavelength, which at optical frequencies may be smaller than the structure length by as much as 20%.

4. Calculation of Dimension

Our numerical results consists of discrete set of frequencies ω_i, $i = 1 \ldots 200$, and the corresponding values of reflection coefficient $R_i(\omega_i)$. These pairs can be interpreted as a set of points on a two-dimensional plane, which can be connected to form a piecewise linear curve, as shown in Figure 5a, where reflectance of the first discussed system for $f = 1/3$ is shown. In other words, consecutive data points connected with negligibly thin lines are interpreted as a two-dimensional, geometrical object in $[\omega, R]$ coordinate space.

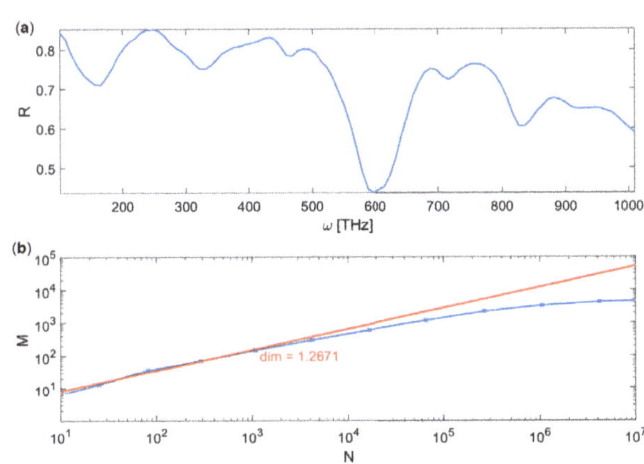

Figure 5. (a) Numerically calculated reflection coefficient for $d = 0$, $h = 45$ nm, $f = 1/3$. (b) The fractal dimension of the curve determined by fitting (red line), according to the Equation (7).

We employ a box-counting method [29,30] to calculate the fractal dimension of this curve. Let's consider a rectangular region of space enclosing the curve and define a grid that divides this space into rectangles (boxes) of a size ζ. The side of the whole grid is $N \sim \zeta^{-1}$ boxes long. Then, we count the number of boxes M which are non-empty, e.g., they contain some part of a curve, which can be either a data point $[\omega_i, R_i]$ or a fragment of a line connecting such points. The so-called Minkowski-Bouligand dimension D is determined by observing how the number of non-empty boxes M scales with their size ζ. For example, for a simple, one-dimensional structure such as infinitely thin straight line, one would obtain $M \sim \zeta^{-1}$ and for two-dimensional area, $M \sim \zeta^{-2}$. One can see that the exponent determines the dimension; therefore, we can use a general expression [29]

$$D = \frac{\partial \log M}{\partial \log N}. \tag{7}$$

Numerically, we estimate the dimension D by plotting the function $\log M(\log N)$ and fitting a straight line, as shown in Figure 5b. One can see that in the limit of $\zeta \rightarrow 0$ (e.g., large number of boxes $N \rightarrow \infty$) the function slope (e.g., dimension) is decreasing. This is caused by our numerical approach, where the curve is represented by a finite number of pixels M_{max}; when the box size approaches the pixel size, the M stops increasing and the curve effectively becomes a set of separate, zero-dimensional points.

The Figure 6a shows the calculated structure and spectrum dimensions D as a function of the fraction f. The numerical results are marked by points; due to the fact that the exact value of the calculated dimension depends on many factors such as the number of points in the spectrum, numerical accuracy and choice of points in the fitting, straight lines are added to the plots to emphasize the general tendencies. We have found that these tendencies are relatively independent of numerical considerations and are preserved when scaling the system. As expected, the dimension of the structure is decreasing with f; one obtains a continuous metal layer at $f = 0$ and a set of separate, point-like islands for large f. The dimension of the reflection spectrum also shows a linear dependence on f, but only for $f < 0.25$. For a very low f, the reflection spectrum is simple, similar to the case of a continuous layer, which results in a low dimension $D \sim 1$. For $f < 0.05$, small gaps between islands cannot be adequately resolved on the numerical grid, leading to spurious results. As the gaps increase, the resulting large metal islands produce multiple, strong plasmonic modes, producing rich reflection spectrum which approaches the asymptotic value of $D \sim 1.25$ for $f \approx 0.25$. Further increase of spacing between islands does not add any new features; as metal islands become smaller, resonant modes shift to higher frequency and eventually become undetectable. This results in a constant spectrum dimension for $f > 0.25$. Figure 6b shows a very similar tendency. However, in this case, it extends to the whole range of f because the spectrum remains relatively feature-rich for all values of f due to the interplay between island and groove modes. As in Figure 1a, the results for structures with small f show the highest variance which is caused by finite numerical accuracy. Due to the overall low reflection coefficient of these structures, the reflection spectrum cannot be calculated reliably enough. Also, the spatial features of structures with small f are comparable in size to the numerical grid, which introduces rounding off errors and aliasing. These factors produce visible, quasi-periodic variations depending on the size of numerical spatial and time steps.

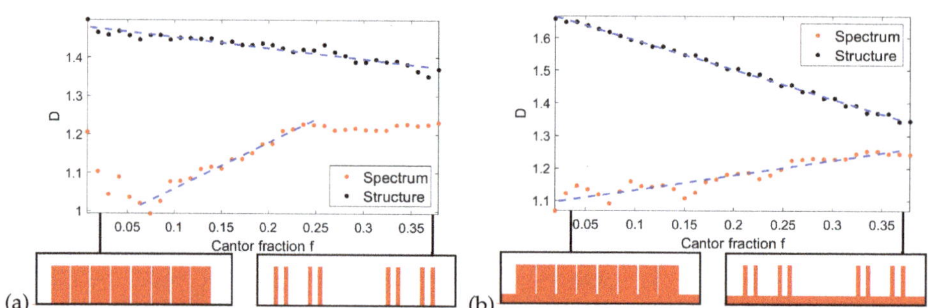

Figure 6. Calculated dimension of the reflection spectrum for (**a**) $d = 0$, $h = 45$ nm and (**b**) $d = 45$ nm, $h = 90$ nm. Insets: schematic of the structure.

Due to the linear dependence of the structure and spectrum dimension on f, one can calculate the correlation between these two quantities. One of the most commonly used measures of correlation is the Pearson correlation coefficient $\rho(x,y) = \frac{cov(x,y)}{\sigma(x)\sigma(y)}$, where $cov(x,y)$ is the covariance of quantities x, y and $\sigma(x)$ is the standard deviation of x [31]. Figure 7a shows a linear fit to the relation between spectrum and structure dimension in the first considered system ($d = 0$, $h = 45$ nm), with $\rho = 0.978$ indicating that there is a strong correlation between these variables. Moreover, a very good fit is obtained for the spectrum dimension and mean groove size (Figure 7b). This result extends also to the island size and effective layer thickness, as these geometrical features are all directly linked—the sum of all islands and grooves is the structure length l and the thickness is a weighted mean of d and $d + h$, where the weight is the ratio of the total island surface to total groove surface. In the discussed case of $d = 0$, the islands are separate metallic structures on glass substrate and the mean groove size is the mean distance between these structures. This means that the results could be applicable to random ensembles of nanoantennas [32–34] and other similar systems [35]. Figure 7c,d depict the same fits performed for the second system ($d = 45$ nm, $h = 90$ nm). The straight lines have the same direction as in the previous case, which indicates that the results are general and repeatable across various systems, despite different absolute values of the structural dimension and spectrum dimension. Finally, we have performed simulation for semi-random structure, where the island and groove sizes are random variables with expected values given by Equation (1) and a standard deviation of 30%. The islands have a height $h = 90$ nm and the thickness of the continuous layer is $d = 45$ nm. Such a system represents a realistic, rough metallic surface. Figure 7e,f shows the correlations obtained in this setup. Due to the complicated geometry with many different distances in the structure, there are overall more plasmonic modes than in previous systems. As a result, the reflection spectrum is richer and its dimension is larger. There is also less variance in the value of dimension—even for small f, the structure can support a large number of plasmonic resonances, so that the spectrum does not become significantly simpler for $f \to 0$.

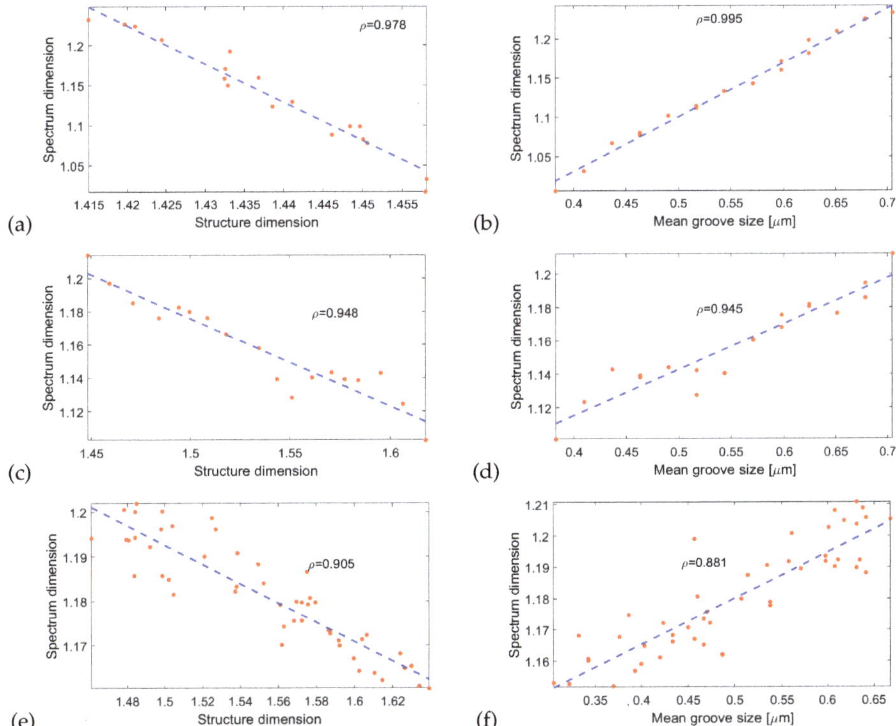

Figure 7. Calculated dimension of the reflection spectrum for (**a**,**b**) $d = 0$, $h = 45$ nm and (**c**–**f**) $d = 45$ nm, $h = 90$ nm, as a function of structure dimension and mean groove size.

To sum up, in both discussed cases the reflection spectra contain a large number of plasmonic modes which can be associated with specific geometrical features of the metallic structure. Due to this dependence, one can correlate the dimension of the structure and the spectrum. This is the key result of this work. The presented relations are repeatable across different systems and hold in the presence of noise. These results suggest that the proposed approach is viable for realistic, random and semi-random fractal-like structures. This facilitates the use of fractal dimension analysis as a useful tool to extract information about the structure from the spectrum, even for the case of a large number of overlapping plasmonic modes. Such correlations could be useful in surface physics [35] because the spectroscopic interpretation of the spectrum enables one to get insight into the material structure of the illuminated system. Moreover, we have shown that the frequencies of plasmonic modes depend on f (Figures 3 and 4), which in turn affects the structure dimension in a linear manner (Figure 6). Therefore, one can associate the dimension with frequency shifts, as reported in [30].

It should be mentioned that apart from f, another degree of freedom in structure generation is the number of Cantor set iterations. We have shown that particular plasmonic modes can be associated with iteration number and therefore the total number of modes is directly dependent on the maximum number of iterations. This is a known effect in superlattices [19]. In studies of light scattering on Cantor set, the intensity spectrum follows iteration-dependent power laws [14].

5. Correlation Between Dimension and Structure Entropy

It has been shown that the Shannon entropy of a structure is closely related to its fractal dimension [15]. In the case of our Cantor set, the structure is numerically represented by a vector of discrete values indicating the thickness of the metal layer along the surface. Specifically, in FDTD calculations we have used $N = 300$ values $d_i, i = 1 \ldots N$, which are either equal to $d_i = d$ or $d_i = d + h$. Therefore, one can conclude that the surface is encoded by N bits of information. To calculate entropy, we use the standard formula [36]

$$S = -\sum_{i=1}^{N} p_i \log_2 p_i, \tag{8}$$

where p_i is the probability of d_i having the given value; for Cantor fraction f, one has

$$p_i = \begin{cases} f, & d_i = d, \\ 1 - f, & d_i = d + h, \end{cases} \tag{9}$$

which results in the total entropy

$$S = -N \left[f^2 \log_2 f + (1 - f)^2 \log_2 (1 - f) \right]. \tag{10}$$

The relation above is a function close to a parabola, with a maximum value of $N/2$ at $f = 0.5$. The comparison of S calculated from Equation (8) and a theoretical relation (10), obtained for the second structure, e.g., $d = 45$ nm, $h = 90$ nm, is shown on the Figure 8a. The results indicate that the system exhibits varying degrees of order, especially in the case of very low or very high f the structure collapses to the trivial case of smooth surface. The peak entropy is half of its maximum possible value, e.g., $S = N$, which would be obtained in a perfectly random structure. As shown before, there is a linear dependence between Cantor fraction f and the structure dimension. Therefore, the entropy as a function of dimension shown in Figure 8b has a similar shape to Figure 8b. Interestingly, the best fit is obtained with a third-degree polynomial and the peak entropy occurs for $D = 1.56$, which is roughly halfway between closest integer dimensions $D = 1$ and $D = 2$. One can conclude that the fractal structure is the richest, in terms of Shannon's information, when the dimension is possibly far from the integer values. Finally, one can express the structure entropy as a function of its reflection spectrum dimension (Figure 8c). Again, a third-degree polynomial with a maximum value at $D = 1.15$ is the best fit and the quality of the fit is better for the more complicated spectra characterized with a larger dimension.

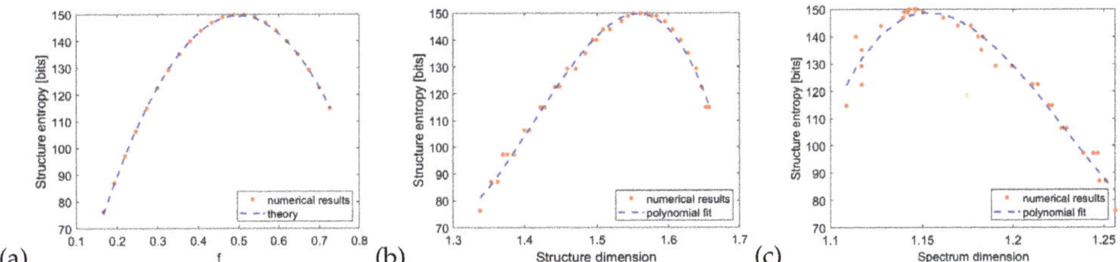

Figure 8. Structure entropy (8) as a function of (**a**) fraction f (**b**) structure dimension (**c**) spectrum dimension.

The entropy calculated with Equation (8) is based on a one-dimensional map of height values, which is a minimal information needed to describe the metallic layer. However, one can also use the direct representation of the structure in the FDTD simulation, e.g., a two-dimensional binary map describing the area with and without metal. In such a case, we have

$$S = -\sum_{i=1}^{N}\sum_{i=1}^{M} p_i \log_2 p_i,$$

(11)

where the area has a size $N \times M$. In such a case, the dependence of entropy on the fraction f becomes linear (Figure 9a). The relation between entropy and dimensions (Figure 9b,c) is also very close to a linear function, which has been proposed in [15], in a more general context of Rényi entropy and generalized fractal dimension. The deviation from the straight line is caused by finite accuracy of box-counting dimension calculation, specifically the choice of points to fit the asymptote shown on Figure 5. This is caused by the fact that the entropy value of a fractal system depends on the scale of measurement (in our case, number of points N), but the fractal dimension is independent of the scales [16]. Finally, from Figure 9c we conclude that it is possible to estimate the structure entropy not only from its fractal dimension but also from the dimension of its reflection spectrum.

It should be stressed that our results indicate that the structure entropy, which is based on the spatial distribution of metallic elements, manifests itself in the reflection spectrum, which is a Fourier transform of the time-dependent signal. This contrasts with a more direct approach such as [17], where the authors have studied the relation between structure entropy and the spatial distribution of reflected wave.

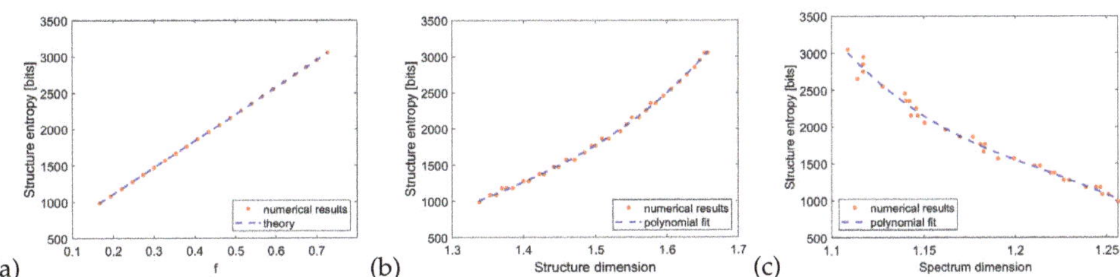

Figure 9. Structure entropy (11) as a function of (**a**) fraction f (**b**) structure dimension (**c**) spectrum dimension.

6. Conclusions

We have shown that a fractal plasmonic system in the form of Cantor set supports multiple standing wave SPP modes which are closely linked to the specific geometrical features of the structure. This dependence allows one to predict the location of extrema in the reflection spectrum. Moreover, it is demonstrated that by interpreting the reflection spectrum as a fractal, one can calculate its non-integer dimension and relate it to the dimension of the illuminated structure. This provides a novel way to extract information about structure from the reflectance. Importantly, the proposed approach based on the spectrum dimension analysis provides clear correlations even in the case where individual resonances cannot be easily resolved. The results could be applied to other systems with non-integer dimensions, including rough-surfaced metallic layers. The presented scheme is based on the well-known Kretschmann configuration and the simple, single-layer system geometry seems to be within reach of the usual fabrication techniques. Experimental data for the discussed problems are not available yet but we hope that our theoretical results will pave the way for future observations and might have practical applications in, e.g., construction of photodetectors, surface-enhanced spectroscopy, photovoltaic devices, and high-gain, compact, multiband antennas [9].

Author Contributions: Conceptualization, D.Z.; data curation, K.K.; investigation, K.K. and S.Z.-R.; methodology, D.Z.; supervision, S.Z.-R.; writing—original draft, D.Z., K.K. and S.Z.-R.

Funding: Support from the National Science Centre, Poland (project OPUS 2017/25/B/ST3/00817) is greatly acknowledged.

Conflicts of Interest: The authors declare no conflict of interest.

References

1. Feis, J.; Mnasri, K.; Khrabustovskyi, A.; Stohrer, C.; Plum, M.; Rockstuhl, C. Surface plasmon polaritons sustained at the interface of a nonlocal metamaterial. *Phys. Rev. B* **2018**, *98*, 115409. [CrossRef]
2. Li, S.; Jadidi, M.M.; Murphy, T.E.; Kumar, G. Terahertz surface plasmon polaritons on a semiconductor surface structured with periodic V-grooves. *Opt. Express* **2013**, *21*, 7041-7049. [CrossRef] [PubMed]
3. Lin, S.; Bhattarai, K.; Zhou, J.; Talbayev, D. Giant THz surface plasmon polariton induced by high-index dielectric metasurface. *Sci Rep.* **2017**, *7*, 9876. [CrossRef] [PubMed]
4. Erementchouk, M.; Joy, S.R.; Mazumder, P. Electrodynamics of spoof plasmons in periodically corrugated waveguides. *Proc. R. Soc. A* **2016**, *472*, 20160616. [CrossRef] [PubMed]
5. Huang, L.; Chen, X.; Bai, B.; Tan, Q.; Jin, G.; Zentgraf, T.; Zhang, S. Helicity dependent directional surface plasmon polariton excitation using a metasurface with interfacial phase discontinuity. *Light Sci. Appl.* **2013**, *2*, e70. [CrossRef]
6. Wen, D.; Yue, F.; Liu, W.; Chen, S.; Chen, X. Geometric metasurfaces for ultrathin optical devices. *Adv. Opt. Mater.* **2018**, 1800348. [CrossRef]
7. de Nicola, F.; Purayil, N.S.P.; Spirito, D.; Miscuglio, M.; Tantussi, F.; Tomadin, A.; de Angelis, F.; Polini, M.; Krahne, R.; Pellegrini, V. Multiband Plasmonic Sierpinski Carpet Fractal Antennas. *ACS Photonics* **2018**, *5*, 2418–2425. [CrossRef]
8. Tang, S.; He, Q.; Xiao, S.; Huang, X.; Zhou, L. Fractal plasmonic metamaterials: physics and applications. *Nanotechnol. Rev.* **2015**. [CrossRef]
9. Wallace, G.Q.; Lagugné-Labarthet, F. Advancements in fractal plasmonics: Structures, optical properties, and applications. *Analyst* **2019**, *144*, 13–30. [CrossRef]
10. Goldstein, S. Random walks and diffusions on fractals. In *Percolation Theory and Ergodic Theory of Infinite Particle Systems*; Kesten, H., Ed.; Springer: New York, NY, USA, 1987.
11. Carpinteri, A.; Sapora, A. Diffusion problems in fractal media defined on Cantor sets. *ZAMM* **2010**, *90*, 203–210. [CrossRef]
12. Berry, M.V. Diffractals. *J. Phys. A* **1979**, *12*, 781–797. [CrossRef]
13. Merlo, D.R.; Martin-Romo, J.A.R.; Alieva, T.; Calvo, M.L. Fresnel diffraction by deterministic fractal gratings: An experimental study. *Opt. Spectrosc.* **2003**, *95*, 131–133. [CrossRef]
14. Cherny, A.Y.; Anitas, E.M.; Kuklin, A.I.; Balasoiu, M.; Osipov, V.A. Scattering from generalized Cantor fractals. *J. Appl. Cryst.* **2010**, *43*, 790–797. [CrossRef]
15. Zmeskal, O.; Dzik, P.; Vesely, M. Entropy of fractal systems. *Comput. Math. Appl.* **2013**, *66*, 135–146. [CrossRef]
16. Chen, Y.; Huang, L. Spatial Measures of Urban Systems: From Entropy to Fractal Dimension. *Entropy* **2018**, *20*, 991. [CrossRef]

17. Cui, T.-J.; Liu, S.; Li, L.-L. Information entropy of coding metasurface. *Light Sci. Appl.* **2016**, *5*, e16172. [CrossRef]
18. Bouazzi, Y.; Soltani, O.; Romdhani, M.; Kanzari, M. Numerical Investigation on the Spectral Properties of One-Dimensional Triadic-Cantor Quasi-Periodic Structure. *Prog. Electromagn. Res. M* **2014**, *36*, 1–7. [CrossRef]
19. Vasconcelos, S.S.; Albuquerque, E.L. Polaritons in Periodic and Quasiperiodic Structures. *Physica B* **1996**, *222*, 113–122. [CrossRef]
20. Khan, S.I.; Islam, S. An Exploration of the Generalized Cantor Set. *Int. J. Sci. Technol. Res.* **2013**, *2*, 7.
21. Novotny, L. Effective Wavelength Scaling for Optical Antennas. *Phys. Rev. Lett.* **2007**, *98*, 266802. [CrossRef]
22. Raether, H. *Surface Plasmons on Smooth and Rough Surfaces and on Gratings*; Springer: New York, NY, USA, 1988; pp. 4–39.
23. Kretschmann, E.; Raether, H. Radiative Decay of Non Radiative Surface Plasmons Excited by Light. *Z. Naturforsch.* **1968**, *23A*, 2135. [CrossRef]
24. Ziemkiewicz, D.; Słowik, K.; Zielińska-Raczyńska, S. Tunable narrowband plasmonics resonances in electromagnetically induced transparency media. *J. Opt. Soc. Am. B* **2017**, *34*, 1981–1988. [CrossRef]
25. Palik, E.D. *Handbook of Optical Constants of Solids*; Academic Press: Cambridge, MA, USA, 1998; Volume 3.
26. Ziemkiewicz, D.; Słowik, K.; Zielińska-Raczyńska, S. Ultraslow long-living plasmons with electromagnetically induced transparency. *Opt. Lett.* **2018**, *43*, 490–493. [CrossRef] [PubMed]
27. Dettmann, C.P.; Frankel, N.E. Potential theory and analytic properties of a Cantor set. *J. Phys A Math. Gen.* **1993**, *26*, 1009–1022. [CrossRef]
28. Okada, N.; Cole, J.B. Effective Permittivity for FDTD Calculation of Plasmonic Materials. *Micromachines* **2012**, *3*, 168–179. [CrossRef]
29. Schroeder, M. *Fractals, Chaos, Power Laws: Minutes from an Infinite Paradise*; W.H.Freeman & Co Ltd.: New York, NY, USA, 1991.
30. Michieli, N.T. Innovative Plasmonic Nanostructures Based on Translation or Scale Invariance for Nano-Photonics. Ph.D. Thesis, Universitá degli Studi de Padova, Padua, Italy, 2014.
31. Dowdy, S.; Wearden, S.; Chilko, D. *Statistics for Research*; Wiley: New York, NY, USA, 2004.
32. Moreau, A.; Ciracı, C.; Mock, J.; Hill, R.T.; Wang, Q.; Wiley, B.J.; Chilkoti, A.; Smith, D.R. Controlled-reflectance surfaces with film-coupled colloidal nanoantennas. *Nature* **2012**, *492*, 11615. [CrossRef]
33. Nazirzadeh, M.; Atar, F.; Turgut, B.; Okyay, A. Random sized plasmonic nanoantennas on Silicon for low-cost broad-band near-infrared photodetection. *Sci. Rep.* **2014**, *4*, 7103. [CrossRef]
34. Nishijima, Y.; Rosa, L.; Juodkazis, S. Surface plasmon resonances in periodic and random patterns of gold nano-disks for broadband light harvesting. *Opt. Express* **2012**, *20*, 11466. [CrossRef]
35. Bozhevolnyi, S.I.; Volkov, V.S.; Leosson, K. Localization and Waveguiding of Surface Plasmon Polaritons in Random Nanostructures. *Phys. Rev. Lett.* **2002**, *89*, 186801. [CrossRef]
36. Lesne, A. Shannon entropy: A rigorous notion at the crossroads between probability, information theory, dynamical systems and statistical physics. *Math. Struct. Comp. Sci.* **2014**, *24*, e240311. [CrossRef]

Article

Interaction and Entanglement of a Pair of Quantum Emitters near a Nanoparticle: Analysis beyond Electric-Dipole Approximation

Miriam Kosik * and Karolina Słowik *

Institute of Physics, Faculty of Physics, Astronomy and Informatics, Nicolaus Copernicus University in Toruń, Grudziadzka 5, 87-100 Torun, Poland
* Correspondence: mkosik@doktorant.umk.pl (M.K.); karolina@fizyka.umk.pl (K.S.)

Received: 18 December 2019; Accepted: 21 January 2020; Published: 23 January 2020

Abstract: In this paper, we study the collective effects which appear as a pair of quantum emitters is positioned in close vicinity to a plasmonic nanoparticle. These effects include multipole–multipole interaction and collective decay, the strengths and rates of which are modified by the presence of the nanoparticle. As a result, entanglement is generated between the quantum emitters, which survives in the stationary state. To evaluate these effects, we exploit the Green's tensor-based quantization scheme in the Markovian limit, taking into account the corrections from light–matter coupling channels higher than the electric dipole. We find these higher-order channels to significantly influence the collective rates and degree of entanglement, and in particular, to qualitatively influence their spatial profiles. Our findings indicate that, apart from quantitatively modifying the results, the higher-order interaction channels may introduce asymmetry into the spatial distribution of the collective response.

Keywords: quantum plasmonics; beyond dipole; entanglement

---

1. Introduction

When subject to resonant illumination, plasmonic nanoparticles are able to focus electromagnetic fields to subwavelength volumes of space [1–3]. Such a tight field confinement is accompanied by a corresponding local field intensity enhancement of up to three orders of magnitude [2]. In quantum plasmonics [4], this effect is usually exploited to boost the interaction strengths between the locally enhanced light and quantum emitters positioned in the hotspots near the nanoparticles. Typically, these quantum emitters are molecules, quantum dots, or crystalline defects. The achieved interaction strengths typically reach the THz regime [5,6], but can be of the order of electron volt [7], outperforming even photonic crystal cavities [8].

These remarkable interaction strengths enable the fast addressing of quantum emitters with light: even in the weak-coupling regime, a quantum transition can occur at timescales of picoseconds, while stationary states are reached within nanoseconds [5,9]. This effect is typically studied in terms of Purcell factors [10], which quantifies the enhancement of the spontaneous emission rate of quantum emitters due to neighboring nanoparticles [11,12].

In this work, we adopt the quantum-optical perspective, according to which a spontaneous emission is a result of a purely quantum origin, arising from a coupling of a quantum emitter to a surrounding electromagnetic field in its vacuum state [13]. The enhancement of a spontaneous emission, however, can be calculated through classical means, as the power enhancement of a source represented by a classical electric dipole, magnetic dipole, or another type of source. The enhancement is conveniently expressed in terms of electromagnetic Green's tensor [14,15]. If multiple quantum emitters are present in the close vicinity of a nanoparticle, they may all couple to the surrounding quantum vacuum. As a result, additional interesting phenomena arise. In particular, the quantum vacuum surrounding the nanoparticle can serve as a carrier for interactions between the emitters [16]; plasmon-enhanced dipole–dipole coupling was studied in [17–19]. This result is derived from the field elimination from the description either in a formalism based on adiabatic elimination of leaky electromagnetic modes [20], or more rigorously, using the electromagnetic Green's tensor-based field quantization in dispersive media and the Markovian approximation [14,19,21–23].

The confinement to subwavelength spatial domains implies that the assumptions of the paradigmatic electric-dipole approximation may break, and higher-order multipolar channels of interaction between matter and light should be taken into account. This is because the electric-dipole approximation is valid if the size-scale of the field modulations is significantly larger than the extent of the quantum emitter, which may not be the case near plasmonic nanoparticles. The significance of higher-order multipolar terms has been suggested [24–29] and verified experimentally [30–34]. Their impact is not only quantitative: the presence of several parallel interaction channels, for example electric and magnetic dipolar or electric quadrupolar, unlocks the possibility of interference [35]. The appealing consequence of destructive interference is that it might lead to spontaneous emission lifetimes enhanced with respect to the free-space values, corresponding to a perspective of linewidths reduced below the "natural level". In the context of realization of quantum information protocols, the increased lifetime might enable quantum information storage in the quantum emitter's excited state for longer times, which might be realized in nanoscale platforms. These effects can be evaluated based on the theory developed in [36].

Naturally, multiple emitters near a nanoparticle could be used to store not only a single excitation per emitter encoding a single quantum bit, but also correlations in the form of quantum entanglement. This effect has been studied before in the plasmonic context [18,20] within the electric-dipole approximation, and has been suggested for the generation of squeezed light [37].

Here, we study how higher-order interaction terms might influence the collective properties of a pair of emitters positioned near a spherical nanoparticle, which belongs to the most typical of geometries investigated in theory and experiments. The studied scenario involves external illumination with a plane wave drive, and the collective properties include the effective inter-emitter coupling, decay rate, and degree of entanglement. We confirm the important impact of higher-order light–matter interaction channels of both a quantitative and a qualitative character.

2. Results

In this section, we introduce the investigated system (Section 2.1) and briefly recapitulate on the theory developed to a large extent in our previous work [36], though extended by the inclusion of classical illumination and studies of entanglement of quantum emitters (Section 2.2). We perform a study of the collective phenomena beyond the electric-dipole channel with an example in the third part of Section 2.3.

2.1. System

The investigated system consists of a pair of two-level quantum emitters, with excited states $|e\rangle_j$ and ground states $|g\rangle_j$, where $j \in \{1, 2\}$ gives the numbers of the emitter. The eigenstates are separated by energy differences $\hbar\omega_j$, where $\hbar$ stands for the reduced Planck constant. Each quantum emitter is described by the set of Pauli operators $\{\sigma_j = |g\rangle_j\langle e|_j, \sigma_j^\dagger\}$. We introduce their extensions to the Hilbert space of the pair of emitters $\Sigma_1 = \sigma_1 \otimes \mathbb{1}$, $\Sigma_2 = \mathbb{1} \otimes \sigma_2$, where $\mathbb{1}$ is an identity operator in the Hilbert space of a given quantum emitter. Our goal is to investigate the stationary entanglement of such a pair of emitters located near a plasmonic nanoparticle whose exemplary geometry is described below. In such a nanoscale setup, a realistic scenario to generate entanglement involves the illumination of the system with an external laser beam. Then, the quantum emitters are coupled to the electromagnetic field of the following electric Fourier components:

$$E(\mathbf{r}, \omega) = \sum_{X=C,V} \mathbf{E}_{\mathrm{inc},X}(\mathbf{r}, \omega) + \mathbf{E}_{\mathrm{scat},X}(\mathbf{r}, \omega). \tag{1}$$

The subscript "inc" stands for the incoming field, which is the illumination which combines a weak laser beam approximated as a classical plane wave (subscript "C" for classical), $\mathbf{E}_{\mathrm{inc},C}(\mathbf{r}, \omega) = \mathbf{E}_{\mathrm{drive}}(\mathbf{r}, \omega_{\mathrm{drive}})\delta(\omega - \omega_{\mathrm{drive}}) + \mathbf{E}^{\star}_{\mathrm{drive}}(\mathbf{r}, \omega_{\mathrm{drive}})\delta(\omega + \omega_{\mathrm{drive}})$, and the background of quantum vacuum fluctuations

$$\mathbf{E}_{\mathrm{inc},V}(\mathbf{r}, \omega) = i\mu_0\omega \int d^3r' \mathbf{G}_0(\mathbf{r}, \mathbf{r}', \omega)\, \mathbf{j}_V(\mathbf{r}', \omega), \tag{2}$$

where μ_0 is the vacuum magnetic permeability and $\mathbf{G}_0(\mathbf{r}, \mathbf{r}', \omega)$ is the electromagnetic Green's tensor of a homogeneous medium connecting the source at a position $\mathbf{r}'$ to the field at a position $\mathbf{r}$.

The subscript "scat" in Equation (1) represents the field scattered at the nanoparticle given by

$$\mathbf{E}_{\mathrm{scat},X}(\mathbf{r}, \omega) = i\mu_0\omega \int d^3r' \mathbf{G}_{\mathrm{scat}}(\mathbf{r}, \mathbf{r}', \omega)\, \mathbf{j}_X(\mathbf{r}', \omega), \tag{3}$$

where $\mathbf{G}_{\mathrm{scat}}(\mathbf{r}, \mathbf{r}', \omega)$ is the electromagnetic Green's tensor representing the scattered part of the electromagnetic field. The source is a current density induced in the nanoparticle either by the classical plane wave ($\mathbf{j}_C(\mathbf{r}', \omega)$) or by the vacuum noise ($\mathbf{j}_V(\mathbf{r}', \omega)$). Naturally, the electric part of the field is accompanied by the magnetic one $\mathbf{B}(\mathbf{r}, \omega) = -\frac{i}{\omega}\nabla \times \mathbf{E}(\mathbf{r}, \omega)$, where i is the imaginary unit. The magnetic field can be decomposed into the incoming and scattered, classical and vacuum-induced components, accordingly.

In general, both the electric and the magnetic components of the field can be coupled to the quantum emitters, i.e., to the electric dipole $\mathbf{d}_j$, magnetic dipole $\mathbf{m}_j$, electric quadrupole $\mathbf{Q}_j$, and higher-order multipolar moments characterizing the quantum transition between the eigenstates. These transition moments are expressed through the matrix elements of the corresponding operators $\mathbf{d}_j = \langle e|\hat{\mathbf{d}}_j|g\rangle$, and similarly for the other multipoles. In this work, we assume the emitters do not support permanent multipolar moments. For more details on multipolar coupling, please see [33,36,38]. Since we work far from the ultrastrong coupling regime, we assume the rotating wave approximation to hold, and apply it in the following interaction Hamiltonian taking into account the electric dipole, magnetic dipole, and electric-quadrupole terms [38]:

$$\mathcal{H}_{\mathrm{int}} = -\sum_j \left\{ \left[\mathbf{E}^-(\mathbf{r}_j)\cdot\mathbf{d}_j + \mathbf{B}^-(\mathbf{r}_j)\cdot\mathbf{m}_j + \nabla\mathbf{E}^-(\mathbf{r}_j):\mathbf{Q}_j\right]\Sigma_j + \Sigma_j^\dagger\left[\mathbf{d}_j^\dagger\cdot\mathbf{E}^+(\mathbf{r}_j) + \mathbf{m}_j^\dagger\cdot\mathbf{B}^+(\mathbf{r}_j) + \mathbf{Q}_j^\dagger:\nabla\mathbf{E}^+(\mathbf{r}_j)\right] \right\}, \tag{4}$$

where in the Schrödinger picture $\mathbf{E}^-(\mathbf{r}) = \int_0^\infty d\omega\mathbf{E}(\mathbf{r}, \omega)$, $\mathbf{E}^+(\mathbf{r}) = [\mathbf{E}^-(\mathbf{r})]^\dagger$, and $\mathbf{E}(\mathbf{r}) = \mathbf{E}^-(\mathbf{r}) + \mathbf{E}^+(\mathbf{r})$. The fields are evaluated at the quantum emitters' positions $\mathbf{r}_j$. In the expression above, the dot $\cdot$ denotes a scalar product, $\nabla\mathbf{E}$ is a dyadic product, and $C : D = \sum_{ij} C_{ij}D_{ji}$ is a double-dot product of tensors C and D.

We would like to now discuss the roles played by different field components in the scenario proposed in this work. Both components, the quantum vacuum and the classical drive, enter the Hamiltonian above.

- The quantum vacuum is a background, playing the role of a carrier of the interactions of the quantum emitters. In open systems like the one considered in this work, this part of the field is tricky to keep track of, since it involves a continuum of optical modes. However, in this case, the quantum vacuum surrounding the emitters can be treated as a reservoir shared between the quantum emitters [16]. Then, it can be eliminated from the evolution equations of the system, leading to an effective picture as given below. The effective picture is obtained in the Markovian approximation, based on the assumption (which is well met in most practical cases) that at considered timescales, the light–matter coupling introduces only a small perturbation to the free dynamics of the field and of the emitter (for rigorous derivations and detailed discussions please see [23,36]). Under the Markovian approximation, one arrives at an effective form of equations describing the evolution of the emitters alone, in which contributions from Lamb shifts δ_j and spontaneous emission rates can be distinguished for individual emitters, and additionally, direct multipole–multipole interactions and collective decay effects arise between multiple emitters. These effects already arise in free space, but can be significantly enhanced and modified in the presence of plasmonic nanoparticles tailoring the properties of the quantum vacuum, quantified with the Green's tensor. Please note that the coupling effect holds even if the frequencies of the emitters are not identical, as long as their difference $|\omega_1 - \omega_2|$ is much smaller than the width of the plasmonic resonance;

- The classical drive is the source of energy in the setup and therefore is necessary to enable stationary entanglement generation. The energy it provides trades off various decay channels described below. Without the drive, the stationary state of the system would be the ground state, which is separable.

2.2. Hamiltonian and Liouvillan

As described in detail in [36], under the Markovian approximation, Equation (4) reduces to the following form of the full Hamiltonian, given here in a frame rotating with the frequency of the driving field ω_{drive}:

$$\mathcal{H}/\hbar = \sum_{j=1,2} \left(\delta\omega_j \Sigma_j^\dagger \Sigma_j + \Omega_j \Sigma_j^\dagger + \Omega_j^\star \Sigma_j \right) + \zeta \Sigma_2^\dagger \Sigma_1 + \zeta^\star \Sigma_1 \Sigma_2^\dagger. \tag{5}$$

Here, $\delta\omega_j = \omega_j + \delta_j - \omega_{\text{drive}}$ is the detuning of the drive from the transition frequency of the jth system corrected by the effective Lamb shift. It corresponds to an effective energy shift of the *j*th emitter. The effective coupling strengths with the classical field are

$$\hbar\Omega_j = -\mathbf{E}_C(\mathbf{r}_j, \omega_{\text{drive}}) \cdot \mathbf{d}_j - \mathbf{B}_C(\mathbf{r}_j, \omega_{\text{drive}}) \cdot \mathbf{m}_j - \left[\nabla \mathbf{E}_C(\mathbf{r}_j, \omega_{\text{drive}}) \right] : \mathbf{Q}_j. \tag{6}$$

They describe interactions with the classical part of the field consisting of the illuminating plane wave and the part scattered at the nanoparticle $\mathbf{E}_C(\mathbf{r}) = \int_0^\infty d\omega \left[\mathbf{E}_{\text{inc},C}(\mathbf{r}, \omega) + \mathbf{E}_{\text{scat},C}(\mathbf{r}, \omega) \right]$ and similarly for the magnetic field.

The final contribution to the Hamiltonian describes the first among the collective effects, i.e., the multipole–multipole interaction of strength ζ. This quantity arises from the presence of the quantum vacuum and is essential for the purpose of entanglement generation since it is the very source of nonclassical correlations of the two emitters. We consider the coupling strength in the form extended with respect to the well-known one corresponding to the electric-dipole approximation (see [17,19] for a result in the electric-dipole approximation and [36] for extensions).

$$\zeta = \pi\bar{\omega}^2 \sum_{mn} \left[R_{mn}\left(\bar{\omega}\right) \text{Re}\, G_{mn}\left(\mathbf{r}', \mathbf{r}, \bar{\omega}\right) |_{\substack{\mathbf{r}=\mathbf{r}_1, \\ \mathbf{r}'=\mathbf{r}_2}} + I_{mn}\left(\bar{\omega}\right) \text{Im}\, G_{mn}\left(\mathbf{r}', \mathbf{r}, \bar{\omega}\right) |_{\substack{\mathbf{r}=\mathbf{r}_1, \\ \mathbf{r}'=\mathbf{r}_2}} \right], \tag{7}$$

where $\bar{\omega} = \frac{1}{2}(\omega_1 + \omega_2)$ and R_{mn}, I_{mn} correspond to the real and imaginary parts of a differential operator acting on the elements of the full Green's tensor $G_{mn} = (G_0 + G_{\text{scat}})_{mn}$,

$$R_{mn}(\omega) = \frac{\mu_0}{\pi\hbar} \text{Re} \left[D_{j,m}^{r'}{}^\dagger(\omega)\, D_{j,n}^{r}(\omega) \right], \quad I_{mn}(\omega) = \frac{\mu_0}{\pi\hbar} \text{Im} \left[D_{j,m}^{r'}{}^\dagger(\omega)\, D_{j,n}^{r}(\omega) \right], \tag{8}$$

with components of the differential operator

$$D_{j,n}^{r}(\omega) = d_{j,n} + \sum_k \left(Q_{j,nk} + \frac{i}{\omega} \sum_p \epsilon_{pkn} m_{j,p} \right) \frac{\partial}{\partial r_k}, \tag{9}$$

and with $n, k, p \in \{x, y, z\}$. Here, $d_{j,n}$ stands for the *n*th spatial component of the transition dipole moment element of the jth quantum emitter, and similarly for the other multipoles. From the structure of Equations (7) and (9), it is clear that the interaction will contain two sorts of terms, i.e., of "pure" and "mixed" origin. To explain their meaning, we consider each quantum emitter as a complicated source combining the electric-dipole, magnetic-dipole, and electric-quadrupole components. Each of these multipolar components is a source of electric and magnetic fields, which are scattered at the nanoparticle. As the other emitter interacts with these scattered fields, we can distinguish

- "Pure" coupling channels, in which the electric-dipole moment of the emitter j interacts with the electric field originating from the electric-dipole moment of the emitter j′, the magnetic-dipole moment of the emitter j is coupled to the magnetic field generated by the magnetic-dipole moment of the emitter j′, and the electric-quadrupole moment of the emitter j- to the modulations of the electric field originating from the electric-quadrupole source corresponding to the emitter j′;
- "Mixed" coupling channels related to interference, for example, an electric-dipole moment of the jth emitter coupled to the electric field generated by the magnetic dipole or electric-quadrupole sources related to the emitter j′, etc.

The same sorts of channels will be distinguished in collective decay and will influence the degree of entanglement.

Please note that in the electric-dipole approximation, the operator in Equation (9) reduces to an element of the electric-dipole moment, and in free space, the expression for ζ reduces to the familiar form $\hbar\zeta = \frac{1}{4\pi\epsilon_0} \frac{\mathbf{d}_j \cdot \mathbf{d}_{j'} - 3\mathbf{d}_j \cdot \hat{\mathbf{r}}_{jj'} \hat{\mathbf{r}}_{jj'} \cdot \mathbf{d}_{j'}}{r_{jj'}^3}$, where $r_{jj'}$ is the distance between the emitters j and j′, while $\hat{\mathbf{r}}_{jj'}$ indicates the direction of a vector connecting them. This form of dipole–dipole coupling has been derived and applied in previous works focusing on electric dipole–dipole interactions in free-space [16] and their modifications near plasmonic nanostructures [17,19]. The same expression corresponds to the Förster resonance energy transfer (FRET) potential [39,40] where a pair of emitters is considered, one of them playing a role of a donor, the other, of an acceptor of a quantum of energy. This simple form is obtained by inserting the free-space Green's function in Equation (7). It may be modified near plasmonic nanostructures influencing the form of the Green's tensor, and as a result, modifying the range of dipole–dipole interactions/FRET [41,42], or due to the broad character of plasmonic resonances, its spectral characteristics. A particular example is related to plasmon—induced resonance energy transfer, in which energy transfer is enabled to acceptors whose transition line is centered at a frequency blue-shifted with respect to the donors [43].

Having mentioned the FRET, we need to explain how it could be possible to achieve a regime of irreversible energy transfer if the Hamiltonian in Equation (7) is Hermitian and describes reversible dynamics. Irreversibility arises naturally from introducing decoherence in the system. In FRET, the decoherence rate dominates over the multipole–multipole coupling ζ by 6 orders of magnitude, and is mostly related to nonradiative processes, e.g., collisions with molecules of a host medium [40]. In general, the shared photonic environment impacts the individual decay rates of the emitters through the famous Purcell effect [10,12], and may induce the corresponding collective rates which describe the sub- and superradiance phenomena in analogy to the Dicke model [16,19,44]. In most cases, decay and decoherence suppress the degree of stationary entanglement unless the subradiant channel is active through which highly entangled but weakly radiating states can be populated.

The individual rates are given by [36]

$$\gamma_{jj} = \frac{2\mu_0}{\hbar}\omega_j^2 \sum_{mn} D_m^{r'\,\dagger}(\omega_j)\, D_n^{r}(\omega_j)\, \mathrm{Im}G_{mn}(\mathbf{r}',\mathbf{r},\omega_j)\,|_{\substack{\mathbf{r}=\mathbf{r}_{j'},\\ \mathbf{r}'=\mathbf{r}_j}}, \qquad (10)$$

and we account for the collective ones through the expression [36]

$$\gamma_{jj'} = 2\pi\bar{\omega}^2 \sum_{mn}\left[R_{mn}(\bar{\omega})\,\mathrm{Im}\,G_{mn}(\mathbf{r}',\mathbf{r},\bar{\omega})\,|_{\substack{\mathbf{r}=\mathbf{r}_{j},\\ \mathbf{r}'=\mathbf{r}_{j'}}} - I_{mn}(\bar{\omega})\,\mathrm{Re}\,G_{mn}(\mathbf{r}',\mathbf{r},\bar{\omega})\,|_{\substack{\mathbf{r}=\mathbf{r}_{j},\\ \mathbf{r}'=\mathbf{r}_{j'}}} \right], \qquad (11)$$

with $j' \neq j$. These rates enter the Liouville term that accounts for the non-Hamiltonian part of the dynamics of the density matrix ρ of the pair of quantum emitters

$$\mathcal{L}(\rho) = \sum_{j,j'=1,2} \mathcal{D}_{\gamma_{jj'}}(\rho, \Sigma_j, \Sigma_{j'}^\dagger), \tag{12}$$

where

$$\mathcal{D}_\gamma(\rho, A, B) = \gamma\left(A\rho B - \frac{1}{2}BA\rho - \frac{1}{2}\rho BA\right). \tag{13}$$

With these tools at hand, one can evaluate the dynamics of the system through the Gorini–Kossakowski–Sudarshan–Lindblad equation [45,46]. However, we are interested to find its stationary solutions ρ which we deem more feasible for experimental investigations. For this purpose, we solve the stationary form of the equation

$$-i\left[\mathcal{H}, \rho\right] + \mathcal{L}(\rho) = 0. \tag{14}$$

Once the stationary density matrix is known, the degree of entanglement between the emitters can be evaluated e.g., in terms of concurrence [47]

$$\mathcal{C}(\rho) = \max\{0, \lambda_1 - \lambda_2 - \lambda_3 - \lambda_4\}, \tag{15}$$

where λ_i stands for square roots of eigenvalues, in a descending order, of the matrix $\rho\tilde{\rho}$, and where $\tilde{\rho} = (\sigma_{y,1} \otimes \sigma_{y,2})\rho^*(\sigma_{y,1} \otimes \sigma_{y,2})$. Here, $\sigma_{y,j} = i(\sigma_j - \sigma_j^\dagger)$.

2.3. Application

The following steps allow one to apply the theory introduced above to an arbitrary reasonable geometry.

- The Green's tensor $\mathbf{G}(\mathbf{r}, \mathbf{r}', \omega)$ corresponding to the particular geometry under study should be found. Here, $\mathbf{r} = \mathbf{r}' = \mathbf{r}_j$ and $\omega = \omega_j$ for single-emitter effects, while $\mathbf{r} = \mathbf{r}_j$, $\mathbf{r}' = \mathbf{r}_{j'}$ and $\omega = \bar{\omega}$ for collective effects, are respectively positions of the probe and the source as well as their frequencies. For this purpose, we have used a freely-available MATLAB solver MNPBEM, as described in the Materials and Methods section;
- Relevant derivatives of the Green's tensor should be evaluated according to the orientations of the multipolar moments (Equation (9)). The first derivatives correspond to the interference terms involving the electric-dipole component, while the second derivatives are related to the magnetic dipole and electric-quadrupole components. Description of higher-order components would require a generalization of the method;
- To evaluate the degree of stationary entanglement in terms of concurrence, one needs to insert the effective-Hamiltonian/Liouvillian parameters, calculated above, to Equation (14) for the stationary density matrix. The concurrence can be found directly according to the recipe in Equation (15). These derivatives scaled by the multipolar moments determine the emitter–emitter interaction strengths and decay rates, according to Formulas (7), (10) and (11).

We now apply this procedure to an example system. To acquire a fair estimation of different multipolar contributions to the degree of entanglement, we assume the dipole moments of each of the emitters to have lengths of 1 atomic unit and the following orientations: $d_x = 1\,\text{a.u.} \approx 8.5 \times 10^{-30}\,\text{Cm}$, $m_z = 1\,\text{a.u} \approx 1.9 \times 10^{-23}\,\text{JT}^{-1}$, and similarly for the electric-quadrupole moment of the transition $Q_{xy} = Q_{yx} = 1\,\text{a.u} \approx 4.5 \times 10^{-40}\,\text{Cm}^2$.

The nanoparticle is a silver nanosphere of 10 nm radius supporting a broad plasmonic resonance centered at a free-space wavelength of 360 nm, whose full-width at half maximum is of the order of 100 nm (not shown). The relative permittivity for silver at 360 nm is $\epsilon = -2.3020 + 0.26535\,i$ [48], the magnetic permeability is assumed to be equal to 1. As we show below, even for such a small nanoparticle, whose optical response is dominated by the electric-dipolar term, the contributions from higher multipoles to the interaction ξ, the decay rates $\gamma_{jj'}$, and to the stationary concurrence

C are considerable. The system is illuminated with a x-polarized plane wave of frequency $\omega_{\text{drive}} = 5.2 \times 10^{15}$ Hz resonant with the nanosphere optical response. This drive is assumed resonant with the Lamb-shift-corrected transition frequency of Emitter 1 $\delta\omega_1 = 0$ and slightly detuned from the other emitter $\delta\omega_2 = 2$ GHz. Please note that this offset corresponds well to possible implementations, in which the Lamb shift correction depends on the emitter's position with respect to the nanoparticle [23]. We fix the position of Emitter 1 at $\mathbf{r}_1 = (-13, 0, 0)$ nm with respect to the coordinate frame's origin at the center of the sphere. We evaluate the resulting model parameters based on the electromagnetic Green's tensor as described in detail in the subsections above. The parameters are calculated in dependence of position $\mathbf{r}_2$ of the other emitter that is swept across the xy plane. The value of the coupling strength with the classical field for Emitter 1 is $\Omega_1 = 1$ GHz, while the couplings Ω_2, ζ and decay rates $\gamma_{jj'}$ are position dependent. Please note that the values of the coupling to the classical field Ω_j only influence our final result: the concurrence, but not the vacuum-induced collective parameters ζ and $\gamma_{jj'}$.

Before we continue to discuss the results, it is important to comment on the case of free space, that is in the absence of the nanoparticle. In free space, the electric-dipole approximation works very well and the light–matter interaction is dominated by the electric-dipole channel. In consequence, the effective parameters, such as the spontaneous emission rate γ_{jj} due to the electric-dipole channel, overcome their analogons due the magnetic dipole and electric-quadrupole channels, respectively, by 5 and 6 orders of magnitude in the studied frequency range [36]. Similar scaling applies to $\gamma_{jj'}$ and ζ. As we show below, this situation may be greatly modified near nanoparticles.

In Figure 1, we compare the resulting coupling strengths $|\zeta|$ achieved due to the electric-dipole interaction channel (a), the magnetic-dipole channel (b), and the electric-quadrupole channel (c). The maps show $|\zeta|$ as functions of the position $\mathbf{r}_2$. The navy-colored regions around the nanoparticle correspond to the positions less than 2 nm apart from the sphere, at which distance our results may not be reliable due to limitations of the software that was used to calculate the Green's tensor [49] and therefore are not shown. We find that all the interaction channels are enhanced, but most importantly, their ratio with respect to the dominant channel is enhanced dramatically. The electric-dipole channel dominates by up to 2 (4) orders of magnitude near the nanoparticle over the magnetic-dipole (electric quadrupole) terms, as is evident from comparisons with Figure 1a–c. This means the relative strength of the higher-order multipoles (i.e., beyond the electric dipole) is enhanced with respect to the free-space case. Even more importantly, the impact of these channels is most visible through the interference terms (Figure 1d–f): The electric–magnetic dipole interference in Figure 1d is the main term responsible not only for the significant modulation of magnitude of the total coupling strength ζ with respect to the value obtained from the isolated electric-dipole channel, but also for the asymmetry with respect to the $y = 0$ plane. We demonstrate this in Figure 2a. There, we find domains where the total interaction strength is robust with respect to positioning of Emitter 2 and modified by interference with the magnetic dipole by $\pm 8\%$ for $\mathbf{r}_2$ below and above the nanosphere, as it is shown in Figure 2e, up to even $\pm 20\%$ for the two emitters located on the same side with respect to the nanoparticle (Figure 2c). If the emitters are positioned on opposite sides, we find the influence of higher-order interaction channels to be limited to a few percent. Please note, however, that even in the latter case, the increase is substantial with respect to the free-space case. One can notice narrow spatial regions where the modulation may exceed 50% (Figure 2c for $y = -14.0$ nm and $y = 12.5$ nm), but these extreme values are achieved in strongly limited volumes and the required experimental accuracy to exploit them would be very challenging to achieve.

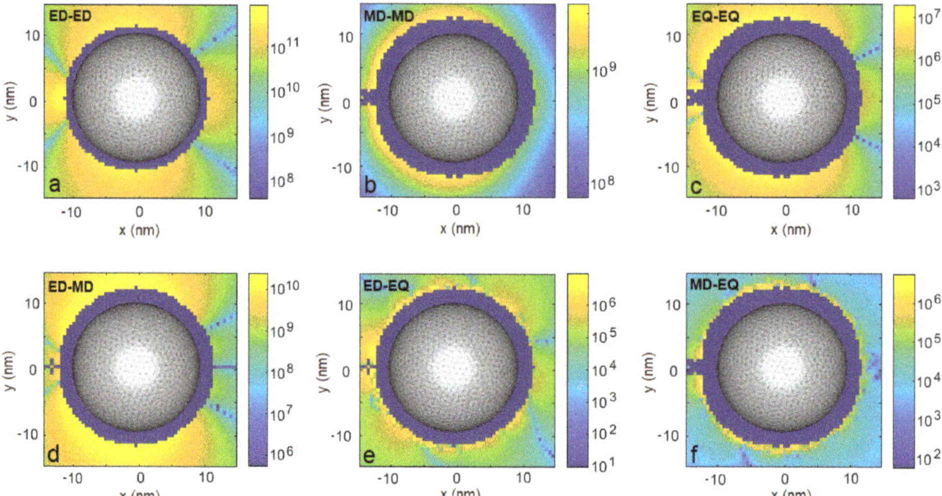

Figure 1. Coupling strengths $|\zeta|$ due to various light–matter interaction channels for Emitter 1 fixed at position $\mathbf{r}_1 = (-13, 0, 0)$ nm as a function of position $\mathbf{r}_2$ of Emitter 2. Orientation of emitters' multipolar moments described in the main text. (**a**). electric dipole–electric dipole coupling, (**b**). magnetic dipole–magnetic dipole coupling, (**c**). electric quadrupole–electric quadrupole coupling, (**d**). electric dipole–magnetic dipole coupling, (**e**). electric dipole–electric quadrupole coupling, (**f**). magnetic dipole–electric quadrupole coupling.

We would like to elaborate on the antisymmetry of the result with respect to the $y = 0$ plane. The physical origin of this effect lies in the interplay of electric- and magnetic-dipole moments of the emitters. The interference term arises as a result of the coupling of the magnetic field induced by the electric dipole corresponding to Emitter 1 and coupled to the magnetic dipole of Emitter 2, as well as the electric field induced by the magnetic dipole of Emitter 1 and coupled to the electric dipole of Emitter 2. The magnetic field induced by the electric dipole of Emitter 1 is dominated by the part scattered at the nanoparticle, which one can imagine as coming from a dipole induced at the nanoparticle. Naturally, the orientation of the magnetic field from such a dipole depends on the position of Emitter 2 which probes it, and in particular has opposite signs in the lower and upper halves of the xy plane.

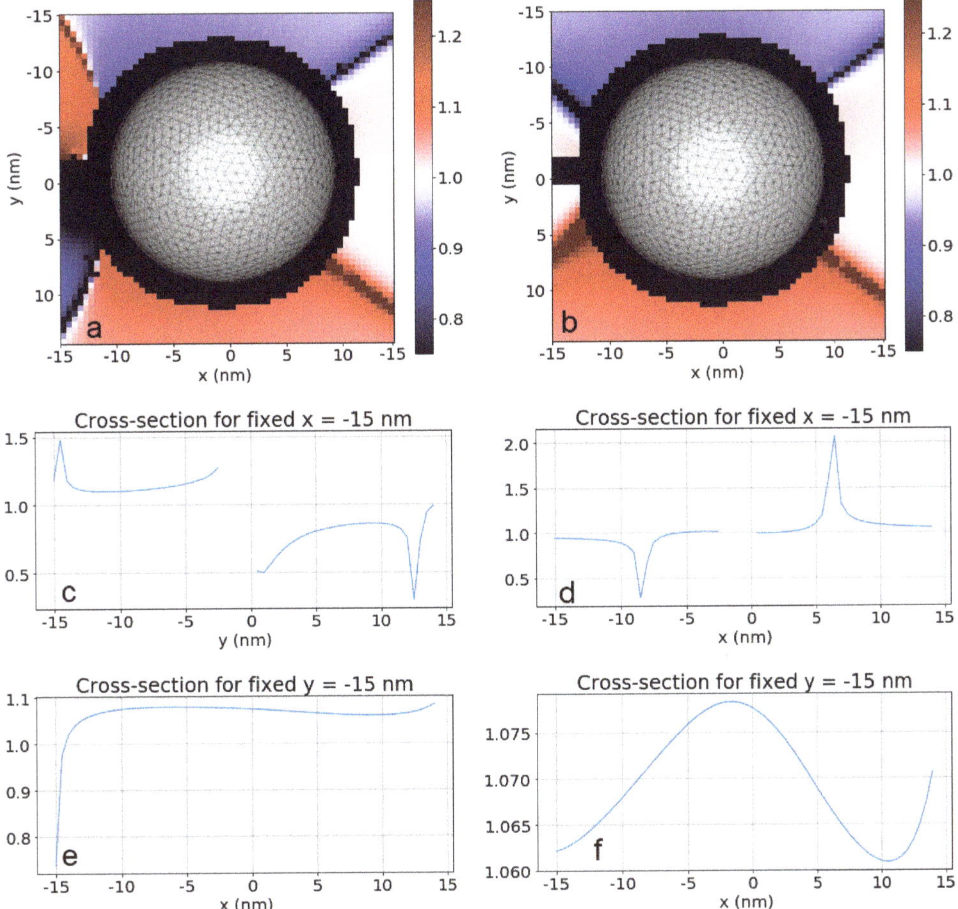

Figure 2. (**a**). Total coupling strength ζ compared to coupling strength ζ_{ED-ED} from the electric-dipole channel only. (**b**). Total emission rate γ_{12} compared to the emission rate $\gamma_{12,ED-ED}$ from the electric-dipole channel only. (**c,d**). Cross-sections of a,b for a fixed $x = -15$ nm. (**e,f**). Cross-sections of a,b for a fixed $y = -15$ nm.

Similar effects can be identified on maps of the collective decay rates γ_{12}, whose absolute values are shown in Figure 3. These rates are rather robust with respect to respective positioning of the emitters: the rates due to the electric-dipole channel reach THz values in almost the entire simulation domain, while those originating from the magnetic-dipole and electric-quadrupole channel are of the order of tens of GHz and hundreds of MHz, respectively. All channels are enhanced for the emitters located very close to each other, i.e., less than a few nanometers apart. Again, the interference between the electric and the magnetic dipoles leads to a substantial modulation of the result (Figure 2d,f) by up to relatively stable values of $\pm 8\%$ in the lower and the upper halves of the xy plane, respectively, and even up to $\pm 100\%$ at specific spots (e.g., at $x = -15$ nm, $y = 6.5$ nm).

Both the multipole–multipole interaction carried by the photonic environment surrounding the nanostructure, and the collective decay rates induced by its presence, give rise to entanglement between the emitters, considerable even in the steady state. The degree of stationary entanglement of the emitters is shown in Figure 4 in terms of stationary concurrence $\mathcal{C}$ again in the function of the position of Emitter 2. Once more, the achieved values are robust against shifts in the emitter positioning: for almost all positions around the nanoparticle, $\mathcal{C}$ is close to its average value $\mathcal{C}_{av} = 0.095$ (the averaging was performed over the investigated and reliable positions only and provides the order of magnitude for the effect). The spatial profile of the concurrence arising as a result of the interplay between all the investigated channels (Figure 4a) inherits the asymmetry that has been found in the coupling strengths and decay rates. As clearly visible from Figure 4b, the origin lies in the inclusion of higher-order interaction channels, in particular the interference between the electric- and magnetic-dipole channels, which influence the spatial profile of the result and modify it quantitatively by inducing the antisymmetry against the $y = 0$ plane.

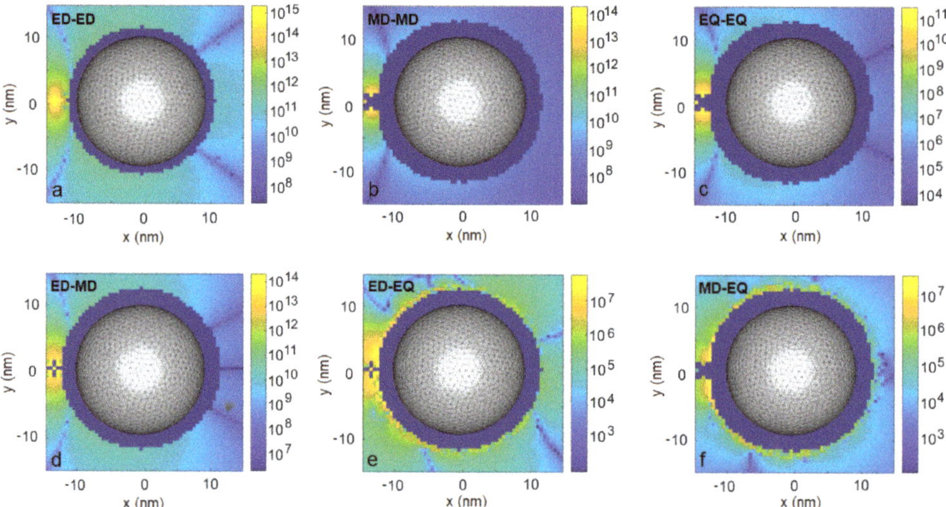

Figure 3. As in Figure 1, but for collective emission rates $|\gamma_{12}|$.

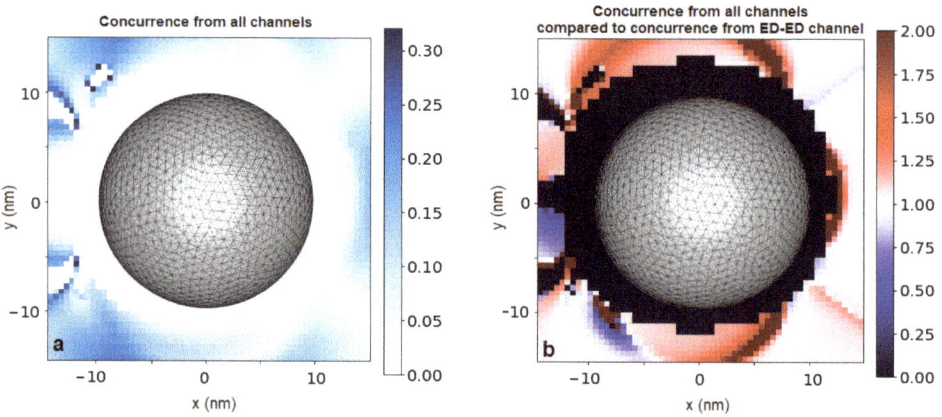

Figure 4. Concurrence due to all possible interaction channels (**a**) and the ratio of concurrence due to all possible channels compared with the concurrence due to electric-dipole channel only (**b**).

To confirm the importance of the plasmonic resonance for the investigated effects, we repeat the calculations of the interaction strengths ξ and collective decay rates $\gamma_{jj'}$ for a drive at free-space wavelength of 350 nm, i.e., blue-shifted to a wing of the plasmonic resonance. The resulting emitter–emitter interactions are not transferred by the nanoparticle to its opposite side as efficiently as in the resonant case (compare Figures 1 and 5). The weaker performance in the off-resonant case can be explained by less efficient field confinement and enhancement, affecting both values of fields and the derivatives, i.e., all considered interaction channels.

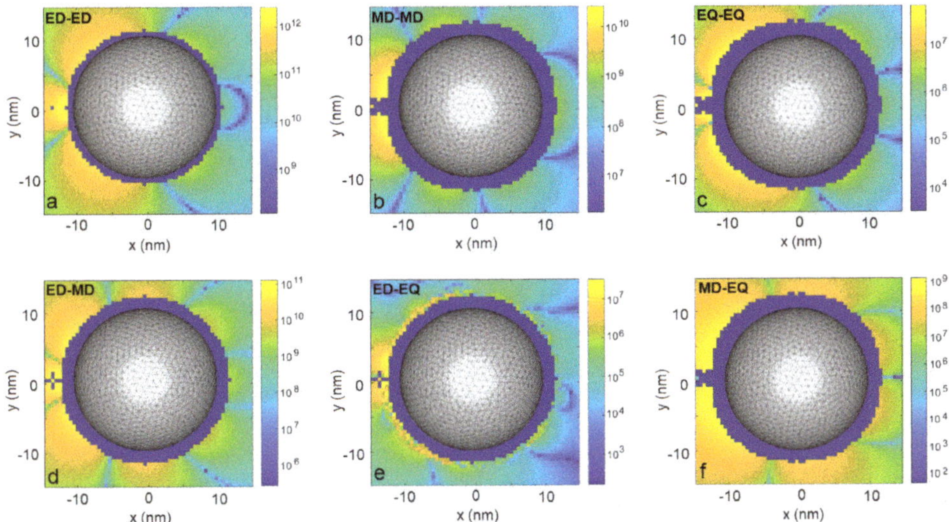

Figure 5. Coupling strengths $|\xi|$ due to various light–matter interaction channels for an illumination wavelength of 350 nm, i.e., detuned from the plasmonic resonance. Emitter 1 is fixed at position $r_1 = (-13, 0, 0)$ nm, coupling strengths are plotted as a function of position r_2 of Emitter 2. (**a**). electric dipole–electric dipole coupling, (**b**). magnetic dipole– magnetic dipole coupling, (**c**). electric quadrupole–electric quadrupole coupling, (**d**). electric dipole–magnetic dipole coupling, (**e**). electric dipole–electric quadrupole coupling, (**f**). magnetic dipole–electric quadrupole coupling.

3. Discussion

Before we come to final conclusions, let us make several comments on the proposed scenario and the acquired results.

The scheme described above to induce multipole–multipole coupling between the emitters and eventually generate entanglement exploits a simple illumination scheme with a plane wave and does not require experimentally challenging preparation of the initial state of the quantum emitters. As demonstrated above, the resulting coupling strengths are rather stable across the simulation domain, which means precise positioning of the emitters is not required. Regarding the orientation of the multipolar moments of the emitters, a certain degree of control would however be beneficial. We have focused our analysis on the orientations in which higher-order interaction channels give rise to spatial asymmetry of the results. We deem this asymmetry to be a measure of the impact of these channels feasible for detection in experimental scenarios. Other orientations lead to similar orders of magnitude for enhancements, but may blur the asymmetry. As long as the dominant components of the transition moments are oriented as assumed above, all the qualitative features of our results should be conserved and significant.

Regarding the degree of entanglement, one could increase its stationary value through refined engineering of the nanoparticle, as well as through the application of a stronger driving field, which would help to overcome the decay rates. In the analysis above, we kept the drive moderate in order not to break the Markovian approximation [20], which is at the heart of the applied model [23].

We analyzed and discussed the influence of light–matter interaction channels beyond the electric dipole on collective phenomena that occur at the presence of a pair of quantum emitters near a silver plasmonic nanosphere. As a result, we found there to be significant influence on these quantities of the interference term between the electric and the magnetic dipoles, which modifies the results obtained within the typically applied electric-dipole approximation. This influence is of both a quantitative and qualitative character. First of all, the interference term may modify the achieved interaction strengths and decay rates by up to 8% in relatively stable spatial domains, and even by 100% at selected points. Secondly, it induces asymmetry of spatial distributions of the investigated quantities. On the other hand, our findings suggest that the influence of even higher-order terms would merely be a small correction, negligible in the particular case of geometry discussed in this work.

In this article, we described methods that could be exploited to include the electric-dipole, magnetic-dipole, and electric-quadrupole light–matter interaction channels in the analysis of quantum-optical scenarios involving small numbers of quantum emitters near nanoparticles. We confirmed the significance of terms beyond the electric-dipole approximation for multipole–multipole coupling, collective decay, and the degree of entanglement of emitters at close vicinity of a silver nanosphere. In the discussed example, we found the dominant correction to originate from the magnetic dipole–electric dipole interference channel. Other geometries where the electric-quadrupole channel may be enhanced were also proposed [50–53] and we expect them to have an impact on the collective effects considered here. Finally, nanostructures made of dielectrics or including dielectric components which support magnetic response at low absorption losses might be used for similar purposes [54–58].

4. Materials and Methods

Detailed introduction to the Green's tensor's formalism for field quantization in dispersive media can be found in [21]. Derivation of effective Hamiltonian parameters beyond the regime of electric-dipole approximation is given in [36].

Calculations of electromagnetic Green's tensor around the investigated nanostructure were performed using the Metallic Nanoparticle Boundary Element Method (MNPBEM) toolbox for Matlab [49]. According to the developers of MNPBEM, to assure reliable results, one should calculate fields at a distance from the particle's surface not smaller than the mean distance between the collocation points of the surface discretization, which in our example is equal to approximately 1 nm. To evaluate the impact of higher-order channels, first and second derivatives of the field are necessary. As a result, the reliable distance from the nanoparticle grows to 2 nm. Scripts are available from the corresponding author upon reasonable request. They can be directly generalized to account for other nanostructure geometries. Concurrence was calculated directly from formula (15) assuming the Markovian approximation holds.

Author Contributions: Conceptualization, K.S.; methodology, M.K. and K.S., software, M.K.; validation, M.K. and K.S.; writing—original draft preparation, K.S.; writing—review and editing, M.K. and K.S.; funding acquisition, K.S. All authors have read and agreed to the published version of the manuscript.

Funding: This research was funded by National Science Centre, Poland grant number 2018/31/D/ST3/01487.

Conflicts of Interest: The authors declare no conflict of interest.

References

1. Fischer, H.; Martin, O.J. Engineering the optical response of plasmonic nanoantennas. *Opt. Express* **2008**, *16*, 9144–9154. [CrossRef] [PubMed]

2. Schuller, J.A.; Barnard, E.S.; Cai, W.; Jun, Y.C.; White, J.S.; Brongersma, M.L. Plasmonics for extreme light concentration and manipulation. *Nat. Mater.* **2010**, *9*, 193. [CrossRef] [PubMed]

3. Gramotnev, D.K.; Bozhevolnyi, S.I. Plasmonics beyond the diffraction limit. *Nat. Photonics* **2010**, *4*, 83. [CrossRef]

4. Tame, M.S.; McEnery, K.; Özdemir, Ş.; Lee, J.; Maier, S.; Kim, M. Quantum plasmonics. *Nat. Phys.* **2013**, *9*, 329–340. [CrossRef]

5. Słowik, K.; Filter, R.; Straubel, J.; Lederer, F.; Rockstuhl, C. Strong coupling of optical nanoantennas and atomic systems. *Phys. Rev. B* **2013**, *88*, 195414. doi:10.1103/PhysRevB.88.195414. [CrossRef]

6. Straubel, J.; Sarniak, R.; Rockstuhl, C.; Słowik, K. Entangled light from bimodal optical nanoantennas. *Phys. Rev. B* **2017**, *95*, 085421. [CrossRef]

7. Chikkaraddy, R.; De Nijs, B.; Benz, F.; Barrow, S.J.; Scherman, O.A.; Rosta, E.; Demetriadou, A.; Fox, P.; Hess, O.; Baumberg, J.J. Single-molecule strong coupling at room temperature in plasmonic nanocavities. *Nature* **2016**, *535*, 127. [CrossRef]

8. Yoshie, T.; Scherer, A.; Hendrickson, J.; Khitrova, G.; Gibbs, H.; Rupper, G.; Ell, C.; Shchekin, O.; Deppe, D. Vacuum Rabi splitting with a single quantum dot in a photonic crystal nanocavity. *Nature* **2004**, *432*, 200. [CrossRef]

9. Barth, M.; Schietinger, S.; Schröder, T.; Aichele, T.; Benson, O. Controlled coupling of NV defect centers to plasmonic and photonic nanostructures. *J. Lumin.* **2010**, *130*, 1628–1634. [CrossRef]

10. Purcell, E. Resonance Absorption by Nuclear Magnetic Moments in a Solid. *Phys. Rev.* **1946**, *69*, 37. [CrossRef]

11. Iwase, H.; Englund, D.; Vučković, J. Analysis of the Purcell effect in photonic and plasmonic crystals with losses. *Opt. Express* **2010**, *18*, 16546–16560. [CrossRef] [PubMed]

12. Akselrod, G.M.; Argyropoulos, C.; Hoang, T.B.; Ciracì, C.; Fang, C.; Huang, J.; Smith, D.R.; Mikkelsen, M.H. Probing the mechanisms of large Purcell enhancement in plasmonic nanoantennas. *Nat. Photonics* **2014**, *8*, 835. [CrossRef]

13. Milonni, P. *The Quantum Vacuum: An Introduction to Quantum Electrodynamics*; Academic Press: Cambridge, MA, USA, 1994.

14. Barnett, S.M.; Huttner, B.; Loudon, R.; Matloob, R. Decay of excited atoms in absorbing dielectrics. *J. Phys. B At. Mol. Opt. Phys.* **1996**, *29*, 3763. [CrossRef]

15. Novotny, L.; Hecht, B. *Principles of Nano-Optics*; Cambridge University Press: Cambridge, UK, 2012.

16. Ficek, Z.; Tanaś, R.; Kielich, S. Quantum beats and superradiant effects in the spontaneous emission from two nonidentical atoms. *Phys. A Stat. Mech. Appl.* **1987**, *146*, 452–482. [CrossRef]

17. Dzsotjan, D.; Kästel, J.; Fleischhauer, M. Dipole-dipole shift of quantum emitters coupled to surface plasmons of a nanowire. *Phys. Rev. B* **2011**, *84*, 075419. [CrossRef]

18. Martin-Cano, D.; González-Tudela, A.; Martín-Moreno, L.; Garcia-Vidal, F.; Tejedor, C.; Moreno, E. Dissipation-driven generation of two-qubit entanglement mediated by plasmonic waveguides. *Phys. Rev. B* **2011**, *84*, 235306. [CrossRef]

19. Sinha, K.; Venkatesh, B.P.; Meystre, P. Collective effects in Casimir-Polder forces. *Phys. Rev. Lett.* **2018**, *121*, 183605. [CrossRef]

20. Hou, J.; Słowik, K.; Lederer, F.; Rockstuhl, C. Dissipation-driven entanglement between qubits mediated by plasmonic nanoantennas. *Phys. Rev. B* **2014**, *89*, 235413. [CrossRef]

21. Dung, H.T.; Knöll, L.; Welsch, D.G. Three-dimensional quantization of the electromagnetic field in dispersive and absorbing inhomogeneous dielectrics. *Phys. Rev. A* **1998**, *57*, 3931. [CrossRef]

22. Scheel, S.; Knöll, L.; Welsch, D.G. Spontaneous decay of an excited atom in an absorbing dielectric. *Phys. Rev. A* **1999**, *60*, 4094. [CrossRef]

23. Dzsotjan, D.; Sørensen, A.S.; Fleischhauer, M. Quantum emitters coupled to surface plasmons of a nanowire: A Green's function approach. *Phys. Rev. B* **2010**, *82*, 075427. [CrossRef]

24. Rolly, B.; Bebey, B.; Bidault, S.; Stout, B.; Bonod, N. Promoting magnetic dipolar transition in trivalent lanthanide ions with lossless Mie resonances. *Phys. Rev. B* **2012**, *85*, 245432. [CrossRef]

25. Hein, S.M.; Giessen, H. Tailoring magnetic dipole emission with plasmonic split-ring resonators. *Phys. Rev. Lett.* **2013**, *111*, 026803. [CrossRef] [PubMed]

26. Tighineanu, P.; Andersen, M.L.; Sørensen, A.S.; Stobbe, S.; Lodahl, P. Probing electric and magnetic vacuum fluctuations with quantum dots. *Phys. Rev. Lett.* **2014**, *113*, 043601. [CrossRef] [PubMed]

27. Rivera, N.; Kaminer, I.; Zhen, B.; Joannopoulos, J.D.; Soljačić, M. Shrinking light to allow forbidden transitions on the atomic scale. *Science* **2016**, *353*, 263–269. [CrossRef] [PubMed]

28. Neuman, T.; Esteban, R.; Casanova, D.; García-Vidal, F.J.; Aizpurua, J. Coupling of molecular emitters and plasmonic cavities beyond the point-dipole approximation. *Nano Lett.* **2018**, *18*, 2358–2364. [CrossRef]

29. Gonçalves, P.; Christensen, T.; Rivera, N.; Jauho, A.P.; Mortensen, N.A.; Soljačić, M. Plasmon-Emitter Interactions at the Nanoscale. *arXiv* **2019**, arXiv:1904.09279.

30. Karaveli, S.; Zia, R. Strong enhancement of magnetic dipole emission in a multilevel electronic system. *Opt. Lett.* **2010**, *35*, 3318–3320. [CrossRef]

31. Taminiau, T.H.; Karaveli, S.; Van Hulst, N.F.; Zia, R. Quantifying the magnetic nature of light emission. *Nat. Commun.* **2012**, *3*, 979. [CrossRef]

32. Aigouy, L.; Cazé, A.; Gredin, P.; Mortier, M.; Carminati, R. Mapping and quantifying electric and magnetic dipole luminescence at the nanoscale. *Phys. Rev. Lett.* **2014**, *113*, 076101. [CrossRef]

33. Kasperczyk, M.; Person, S.; Ananias, D.; Carlos, L.D.; Novotny, L. Excitation of magnetic dipole transitions at optical frequencies. *Phys. Rev. Lett.* **2015**, *114*, 163903. [CrossRef] [PubMed]

34. Li, D.; Karaveli, S.; Cueff, S.; Li, W.; Zia, R. Probing the Combined Electromagnetic Local Density of Optical States with Quantum Emitters Supporting Strong Electric and Magnetic Transitions. *Phys. Rev. Lett.* **2018**, *121*, 227403. [CrossRef] [PubMed]

35. Rusak, E.; Straubel, J.; Gładysz, P.; Göddel, M.; Kędziorski, A.; Kühn, M.; Weigend, F.; Rockstuhl, C.; Słowik, K. Tailoring the Enhancement of and Interference among Higher Order Multipole Transitions in Molecules with a Plasmonic Nanoantenna. *arXiv* **2019**, arXiv:1905.08482.

36. Kosik, M.; Burlayenko, O.; Fernandez-Corbaton, I.; Rockstuhl, C.; Słowik, K. Interaction of atomic systems with quantum vacuum beyond electric dipole approximation. *arXiv* **2019**, arXiv:1911.06166.

37. Haakh, H.R.; Martín-Cano, D. Squeezed light from entangled nonidentical emitters via nanophotonic environments. *ACS Photonics* **2015**, *2*, 1686–1691. [CrossRef]

38. Barron, L.D.; Gray, C.G. The multipole interaction Hamiltonian for time dependent fields. *J. Phys. A Math. Nucl. Gen.* **1973**, *6*, 59. [CrossRef]

39. Masters, B. Paths to Förster's resonance energy transfer (FRET) theory. *Eur. Phys. J. H* **2014**, *39*, 87–139. [CrossRef]

40. Nelson, P.C. The role of quantum decoherence in FRET. *Biophys. J.* **2018**, *115*, 167–172. [CrossRef]

41. Ferri, C.; Inman, R.; Rich, B.; Gopinathan, A.; Khine, M.; Ghosh, S. Plasmon-induced enhancement of intra-ensemble FRET in quantum dots on wrinkled thin films. *Opt. Mater. Express* **2013**, *3*, 383–389. [CrossRef]

42. Zhang, X.; Marocico, C.A.; Lunz, M.; Gerard, V.A.; Gun'ko, Y.K.; Lesnyak, V.; Gaponik, N.; Susha, A.S.; Rogach, A.L.; Bradley, A.L. Experimental and theoretical investigation of the distance dependence of localized surface plasmon coupled Forster resonance energy transfer. *ACS Nano* **2014**, *8*, 1273–1283. [CrossRef]

43. Li, J.; Cushing, S.K.; Meng, F.; Senty, T.R.; Bristow, A.D.; Wu, N. Plasmon-induced resonance energy transfer for solar energy conversion. *Nat. Photonics* **2015**, *9*, 601. [CrossRef]

44. Dicke, R.H. Coherence in spontaneous radiation processes. *Phys. Rev.* **1954**, *93*, 99. [CrossRef]

45. Lindblad, G. On the generators of quantum dynamical semigroups. *Commun. Math. Phys.* **1976**, *48*, 119–130. [CrossRef]

46. Gorini, V.; Kossakowski, A.; Sudarshan, E.C.G. Completely positive dynamical semigroups of N-level systems. *J. Math. Phys.* **1976**, *17*, 821–825. [CrossRef]

47. Hill, S.; Wootters, W.K. Entanglement of a pair of quantum bits. *Phys. Rev. Lett.* **1997**, *78*, 5022. [CrossRef]

48. Johnson, P.B.; Christy, R.W. Optical Constants of the Noble Metals. *Phys. Rev. B* **1972**, *6*, 4370–4379. doi:10.1103/PhysRevB.6.4370. [CrossRef]

49. Hohenester, U.; Trügler, A. MNPBEM–A Matlab toolbox for the simulation of plasmonic nanoparticles. *Comput. Phys. Commun.* **2012**, *183*, 370–381. [CrossRef]

50. Kern, A.; Martin, O.J. Strong enhancement of forbidden atomic transitions using plasmonic nanostructures. *Phys. Rev. A* **2012**, *85*, 022501. [CrossRef]

51. Filter, R.; Mühlig, S.; Eichelkraut, T.; Rockstuhl, C.; Lederer, F. Controlling the dynamics of quantum mechanical systems sustaining dipole-forbidden transitions via optical nanoantennas. *Phys. Rev. B* **2012**, *86*, 035404. [CrossRef]

52. Yannopapas, V.; Paspalakis, E. Giant enhancement of dipole-forbidden transitions via lattices of plasmonic nanoparticles. *J. Mod. Opt.* **2015**, *62*, 1435–1441. [CrossRef]

53. Wu, T.; Zhang, W.; Wang, R.; Zhang, X. A giant chiroptical effect caused by the electric quadrupole. *Nanoscale* **2017**, *9*, 5110–5118. [CrossRef]

54. Schmidt, M.K.; Esteban, R.; Sáenz, J.; Suárez-Lacalle, I.; Mackowski, S.; Aizpurua, J. Dielectric antennas-a suitable platform for controlling magnetic dipolar emission. *Opt. Express* **2012**, *20*, 13636–13650. [CrossRef] [PubMed]
55. Staude, I.; Miroshnichenko, A.E.; Decker, M.; Fofang, N.T.; Liu, S.; Gonzales, E.; Dominguez, J.; Luk, T.S.; Neshev, D.N.; Brener, I.; et al. Tailoring directional scattering through magnetic and electric resonances in subwavelength silicon nanodisks. *ACS Nano* **2013**, *7*, 7824–7832. [CrossRef] [PubMed]
56. Bakker, R.M.; Permyakov, D.; Yu, Y.F.; Markovich, D.; Paniagua-Domínguez, R.; Gonzaga, L.; Samusev, A.; Kivshar, Y.; Luk'yanchuk, B.; Kuznetsov, A.I. Magnetic and electric hotspots with silicon nanodimers. *Nano Lett.* **2015**, *15*, 2137–2142. [CrossRef] [PubMed]
57. Guo, R.; Rusak, E.; Staude, I.; Dominguez, J.; Decker, M.; Rockstuhl, C.; Brener, I.; Neshev, D.N.; Kivshar, Y.S. Multipolar coupling in hybrid metal–dielectric metasurfaces. *ACS Photonics* **2016**, *3*, 349–353. [CrossRef]
58. Kuznetsov, A.I.; Miroshnichenko, A.E.; Brongersma, M.L.; Kivshar, Y.S.; Luk'yanchuk, B. Optically resonant dielectric nanostructures. *Science* **2016**, *354*, aag2472. [CrossRef]

Article

Entropy and Information Theory: Uses and Misuses

Arieh Ben-Naim

Department of Physical Chemistry, The Hebrew University of Jerusalem, Edmond J. Safra Campus Givat Ram, Jerusalem 91904, Israel; ariehbennaim@gmail.com

Received: 7 November 2019; Accepted: 28 November 2019; Published: 29 November 2019

Abstract: This article is about the profound misuses, misunderstanding, misinterpretations and misapplications of entropy, the Second Law of Thermodynamics and Information Theory. It is the story of the "Greatest Blunder Ever in the History of Science". It is not about a single blunder admitted by a single person (e.g., Albert Einstein allegedly said in connection with the cosmological constant, that this was his greatest blunder), but rather a blunder of gargantuan proportions whose claws have permeated all branches of science; from thermodynamics, cosmology, biology, psychology, sociology and much more.

Keywords: entropy; second law; thermodynamics; Shannon measure of information; information theory

---

1. Introduction

Einstein on Thermodynamics:

"It is the only physical theory of universal content, which I am convinced, that within the framework of applicability of its basic concepts will never be overthrown."

Einstein on Infinities:

"Two things are infinite: the universe and human stupidity; and I'm not sure about the universe."

Louis Essen cautions:

"I was warned that if I persisted [criticizing relativity], I was likely to spoil my career prospects."

Adding that:

"Students are told that the theory must be accepted although they cannot expect to understand it. They are encouraged right at the beginning of their careers to forsake science in favor of dogma."

This article tells the incredible story of a simple, well-defined quantity having well-defined limits of applicability, which had evolved to reach monstrous proportions, embracing and controlling everything that happens, and explaining everything that is unexplainable.

My main purpose is to deal with the following perplexing question and quandary: How, a simple, well-defined, "innocent" and powerless concept, called inadequately "entropy" had evolved into an almighty, omnipresent, omnipotent, multi-meaning and monstrous entity, rendering it almost God-like which drives everything and is the cause of everything that happens.

Did I say God-like? Sorry, that was an understatement. Entropy has become far more powerful than any God I can imagine!

God, as we all learned in school, is indeed the almighty and the creator of everything, including us. Yet, we also learn that God gave us the freedom to think, to feel, and even the freedom of believing in Him.

Entropy, in contrast has been ascribed powers far beyond those ascribed to God. It not only drives the entire universe, but also our thoughts, feelings and even creativity.

If you think I am exaggerating in saying that entropy has become mightier than God, please read the following quotation from Michael Shermer's book "Heavens on Earth" [1]:

> *This leads us to a deeper question: Why do we have to die at all? Why couldn't God or Nature endow us with immortality? The answer has to do with ... the Second Law of Thermodynamics, or the fact that there's an arrow of time in our universe that leads to entropy and the running down of everything.*

These statements are typical nonsenses written in many popular science books. Entropy is viewed not only as the *cause* of a specific process, but a cause for anything that happens including our thinking, feelings and creation of the arts [2,3]. I believe that such a claim is the most perverted view of entropy and the Second Law.

1.1. The Unique Nature of Entropy

Entropy is a unique quantity, not only in thermodynamics but perhaps in the entire science. It is unique in the following sense:

There is no other concept in science which has been given so many interpretations—all, except one being totally unjustified. Surprisingly, the number of such misinterpretations still balloons to this day.

(1) There is no other concept which was gravely misused, misapplied and in some instances, even abused.

(2) There is no other concept to which so many powers were ascribed, none of which was ever justified.

(3) There is no other concept which features explicitly in so many book's covers and titles. And ironically some of these book though mentioning entropy, are, in fact totally irrelevant to entropy.

(4) There is no other concept which was once misinterpreted, then the same interpretation was blindly and uncritically propagated in the literature by so many scientists.

(5) There is no other concept on which people wrote whole books full of "whatever came to their mind," knowingly or unknowingly that they shall be immune from being proven wrong.

(6) There is no other concept in physics which was *Equated to time* (not only by words, but by using equality sign "="), in spite of the fact that it is totally timeless quantity [4,5].

(7) There is no other concept in which the role of "cause" and "effect" have been interchanged.

(8) Finally, I would like to add my own "no-other-concept" that has contributed more to the confusion, distortion and misleading the human minds than entropy and the Second Law.

If you doubt the veracity of my statement above, consider the following quotation from Atkins' book [2]:

> *" ... no other scientific law has contributed more to the liberation of the human spirit than the Second Law of thermodynamics."*

The truth is much different:

> *No other scientific law has liberated the spirit of some scientists to say whatever comes to their minds on the Second Law of thermodynamics!*

We shall present *many* similar meaningless and misleading statements about entropy and the Second Law throughout this article.

1.2. Outline of the Article

The article is organized in four Sections, as follows: In Section 2 we describe, very briefly, three different definitions of entropy which are equivalent. These definitions are different in the sense that they are derived from completely different origins, and there is no general proof, that they are identical (i.e., that one follows from the other). Nevertheless, they are equivalent in the sense that for whatever process for which we can calculate the entropy changes we obtain agreement between the results obtained by the different definitions.

In Section 3 we present a few formulations of the Second Law. Some of these are "classical" and one is relatively new. We shall present one thermodynamic and one probability formulation of the Second Law.

In Section 4 we discuss some of the most common interpretations of entropy. We shall see that none of these, except one is a valid and provable interpretation.

In Section 5 we shall discuss three directions along which entropy and the Second Law were misused and misapplied. In this Section we discuss the heart of the greatest blunder in the history of science.

The first is the association of entropy with time. The second is the application of entropy to life phenomena, to evolution to social processes and beyond. The third is the application of entropy and the Second Law to the entire universe; not only at present but also at some hypothetical time in the far past (the "birth" of the universe), or in the far future (the so-called thermal-death of the universe).

This paper is essentially a transcription of an invited lecture the author gave in a conference on Thermodynamics and Information Theory, organized by Professor Adam Gadomski, and held in Bydgoszcz, Poland, October 2019. All these topics are discussed in greater details in Ben-Naim [4–9].

2. Three Different but Equivalent Definitions of Entropy

Here we shall very briefly present three different definitions of Entropy, with more details available in Ben-Naim (henceforth ABN) [4–9]. By "different" we mean that the definitions do not follow from each other, specifically, neither Boltzmann's nor ABN's definition can be "derived" from Clausius's definition. By "equivalent" we mean that for all processes for which we can calculate their change in entropy, we obtain the same results by using the three definitions. It is believed that these three definitions are indeed equivalent although no formal proof of this is available.

2.1. Clausius's "Sefinition" of Entropy

Clausius did not really *define* entropy. Instead, he defined a small change in entropy for one very specific process. It should be added that even before a proper definition of entropy was offered, it was realized that entropy is a *state function*. That means that whenever the *macroscopic state* of a system is specified, its entropy is determined. Clausius' definition, together with the Third Law of Thermodynamics, led to the calculation of "absolute values" of the entropy of many substances.

Let $dQ > 0$ be a *small quantity* of heat flowing *into* a system, being at a given temperature T. The *change* in *entropy* is defined as:

$$dS = \frac{dQ}{T} \tag{1}$$

The letter d here stands for a very *small quantity*, and T is the absolute temperature. Q, has the units of *energy*, and T has the units of *temperature*. Therefore, the entropy change has the units of *energy* divided by units of *temperature*. Sometimes, you might find the subscript "rev" in the Clausius definition, meaning, that Equation (1) is valid only for a "reversible." This is not necessary; it is sufficient to state that the heat is added in a quasi-static process [4].

The quantity of heat, dQ. *must* be very small, such that when it is transferred into, or out from the system, the temperature T does not change. If dQ is a finite quantity of heat, and one transfers it to a system which is initially at a given T, the temperature of the system might change, and therefore the change in entropy will depend on both the initial and the final temperature of the system. There are many processes which do not involve heat transfer, yet, from Clausius' definition, and the postulate that the entropy is a *state function*, one could devise a *path* leading from one state to another, for which the entropy change can be calculated [4].

2.2. Boltzmann's Definition Based on Total Number of Micro-States

Boltzmann defined the entropy in terms of the *total of number accessible micro-states* of a system consisting of a huge number of particles, but characterized by the macroscopic parameters of energy E, volume V and number of particles N.

Consider a gas consisting of N simple particles in a volume V, each particle's micro-state may be described by its location vector R_i and its velocity vector v_i, Figure 1. By simple particles we mean particles having no internal degrees of freedom. Atoms such as argon, neon and the like are considered as simple. They all have internal degrees of freedom, but these are assumed to be unchanged in all the processes we discuss here. Assuming that the gas is very dilute so that interactions between the particles can be neglected, then, all the energy of the system is simply the sum of the kinetic energies of all the particles.

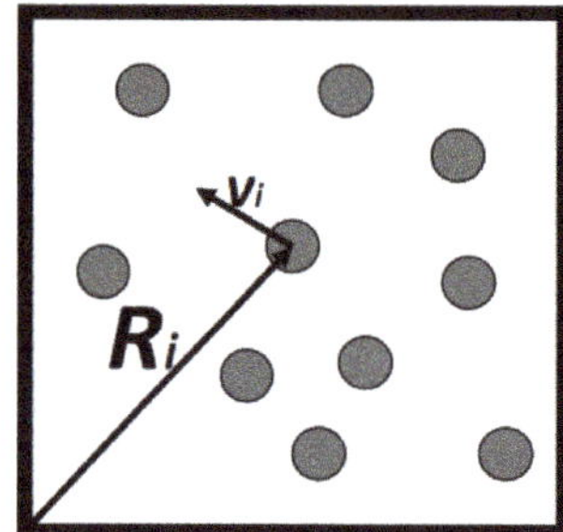

Figure 1. Ten particles in a box of volume V. Each particle, i has a locational and a velocity vector.

Boltzmann postulated the relationship which is now known as the Boltzmann entropy.

$$S = k_B \log W \tag{2}$$

where k_B is a constant, now known as the Boltzmann constant (1.380×10^{-23} J/K), and W is the number of accessible micro-states of the system. Here, log is the natural logarithm. At first glance, Boltzmann's entropy seems to be completely different from Clausius' entropy. Nevertheless, in all cases for which one can calculate changes of entropy one obtains agreements between the values calculated by the two methods. Boltzmann's entropy, as defined in Equation (2), has raised considerable confusion regarding the question of whether entropy is, or isn't a subjective quantity [4].

One example of this confusion which features in many popular science books is the following: Entropy is assumed to be related to our knowledge of the state of the system. If we "know" that the system is at some specific state, then the entropy is zero. Thus, it seems that the entropy is dependent on whether one knows or does not know in which state (or states) the system is.

This confusion arose from misunderstanding W which is the *total* number of accessible micro-states of the system. If $W = 1$, then the entropy of the system is indeed zero (as it is for many substances at 0 K). However, if there are W states and we know in which state the system is, the entropy is still $k \ln W$ and not zero!

In general, the Boltzmann entropy does not provide an explicit entropy function. However, for some specific systems one can derive an entropy function, based on Boltzmann's definition. The most famous case is the entropy of an ideal gas, for which one can derive an explicit entropy function. This function was derived by Sackur [10] and by Tetrode [11] in 1912, by using the Boltzmann definition of entropy. We shall derive this function based on Shannon's measure of information in the following section.

2.3. ABN's Definition of Entropy Based on Shannon's Measure of Information

The third definition which I will refer to as the ABN definition. This is Ben-Naim's definition based on Shannon's measure of information (SMI). The reader who is not familiar with SMI is referred to [7]. Here we present only a brief account of Shannon motivation, and his definition of SMI, then we outline the definition of entropy based on SMI.

Shannon was interested in a theory of *communication of information*, not information itself. This is very clear to anyone who read through Shannon's original article [12]. In fact, in the introduction to the book, we find:

> *"The word information, in this theory, is used in a special sense that must not be confused with its ordinary usage. In particular, information must not be confused with meaning."*

Here is how Shannon introduced the measure of Information:

> *Suppose we have a set of possible events whose probabilities of occurrence are $p_1, p_2, \cdots, p_n$. These probabilities are known but that is all we know concerning which event will occur. Can we find a measure of how much "choice" is involved in the selection of the event or how uncertain we are of the outcome?*

If there is such a measure, say, $H\ (p_1, p_2, \ldots, p_n)$, it is reasonable to require of it the following properties:

(1)   *H should be continuous in the p_i.*
(2)   *If all the p_i are equal, $p_i = \frac{1}{n}$, then H should be a monotonic increasing function of n. With equally likely events there is more choice, or uncertainty, when there are more possible events.*
(3)   *If a choice be broken down into two successive choices, the original H should be the weighted sum of the individual values of H.*

Then Shannon proved the theorem: The only H satisfying the three assumptions above has the form:

$$H = -K \sum p_i \log p_i \tag{3}$$

In the following we shall briefly outline the derivation of the entropy of an ideal gas from the SMI. Details may be found in Ben-Naim, Reference [7]. K, in the Shannon article is any constant. In application to thermodynamics K turns into Boltzmann Constant.

2.3.1. First Step: The Locational SMI of a Particle in a 1D Box of Length L

Suppose we have a particle confined to a one-dimensional (1D) "box" of length L. We can define, as Shannon did, the following quantity by analogy with the discrete case:

$$H[f(x)] = - \int f(x) \log f(x) dx \tag{4}$$

It is easy to calculate the density distribution, $f(x)$ which maximizes the locational SMI, $H[f(x)]$ in Equation (4). The result is:

$$f_{eq}(x) = \frac{1}{L} \tag{5}$$

This is a uniform distribution, Figure 2.

$$H(\text{locations in } 1D) = \log L \tag{6}$$

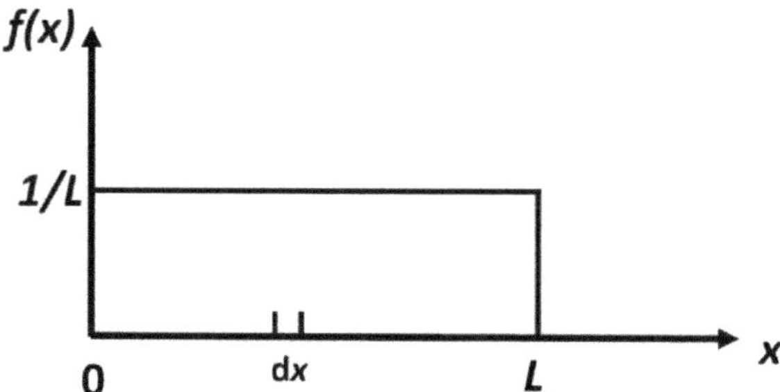

Figure 2. The uniform distribution for a particle in a 1D box of length L. The probability of finding a particle in a small interval dx, is dx/L.

2.3.2. Second Step: The Velocity SMI of a Particle in a 1D "Box" of Length L

The mathematical problem is to calculate the probability distribution that maximizes the continuous SMI, subject to two conditions, which are essentially a normalization condition and a constant variation. The result is the normal distribution, see Figure 3:

$$f_{eq}(x) = \frac{\exp\left[-x^2/2\sigma^2\right]}{\sqrt{2\pi\sigma^2}} \tag{7}$$

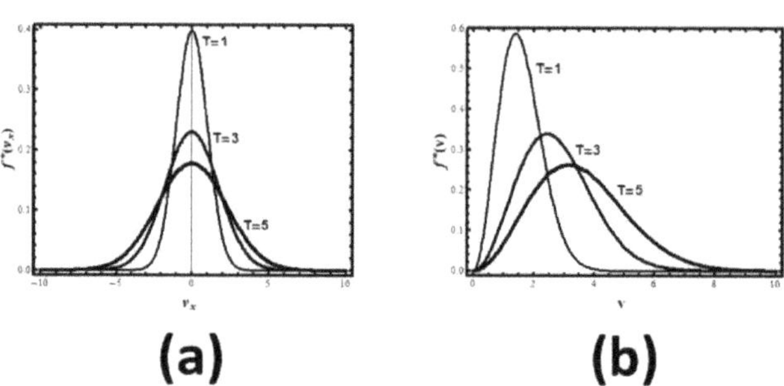

Figure 3. (**a**) The velocity distribution of particles in one dimension at different temperatures; (**b**) The speed (or absolute velocity) distribution of particles in 3D at different temperatures.

The subscript *eq* which stands for *equilibrium* is clarified once we realize that this is the equilibrium distribution of velocities. Applying this result to a classical particle having average kinetic energy $\frac{m<v_x^2>}{2}$, and using the relationship between the standard deviation σ^2 and the temperature of the system:

$$\sigma^2 = \frac{k_B T}{m} \tag{8}$$

We get the equilibrium velocity distribution of one particle in a 1D system:

$$f_{eq}(v_x) = \sqrt{\frac{m}{2\pi k_B T}} \exp\left[\frac{-mv_x^2}{2k_B T}\right] \tag{9}$$

where k_B is the Boltzmann constant, m is the mass of the particle, and T the absolute temperature. The value of the continuous SMI for this probability density is:

$$H_{max}(velocity\ in\ 1D) = \frac{1}{2}\log(2\pi e k_B T/m) \tag{10}$$

Similarly, we can write the momentum distribution in 1D, by transforming from $v_x \rightarrow p_x = mv_x$, to get:

$$f_{eq}(p_x) = \frac{1}{\sqrt{2\pi mk_BT}} \exp\left[\frac{-p_x^2}{2mk_BT}\right] \tag{11}$$

and the corresponding maximum SMI:

$$H_{max}(momentum\ in\ 1\ D) = \frac{1}{2}\log(2\pi emk_BT) \tag{12}$$

2.3.3. Third Step: Combining the SMI for the Location and Momentum of One Particle; Introducing the Uncertainty Principle

We now combine the two results. Assuming that the location and the momentum (or velocity) of the particles are independent events we write:

$$H_{max}(location\ and\ momentum) = H_{max}(location) + H_{max}(momentum) = \log\left[\frac{L\sqrt{2\pi emk_BT}}{h_xh_p}\right] \tag{13}$$

Here h_x and h_p are chosen to eliminate the divergence of the SMI. In writing (14) we assume that the location and the momentum of the particle are independent. However, quantum mechanics impose restrictions on the accuracy in determining both the location x and the corresponding momentum p_x. We must acknowledge that nature imposes on us a limit on the accuracy with which we can determine *simultaneously* the location and the corresponding momentum. Thus, in Equation (14), h_x and h_p cannot both be arbitrarily small, but their product must be of the order of Planck constant $h = 6.626 \times 10^{-34}$ J s. Thus, we set:

$h_xh_p \approx h$, and instead of (14), we write:

$$H_{max}(location\ and\ momentum) = \log\left[\frac{L\sqrt{2\pi emk_BT}}{h}\right] \tag{14}$$

2.3.4. The SMI of a Particle in a Box of Volume V

We consider again one simple particle in a cubic box of volume V. We assume that the location of the particle along the three axes x, y and z are independent. Therefore, we can write the SMI of the location of the particle in a cube of edges L, and volume V as:

$$H(location\ in\ 3D) = 3H_{max}(location\ in\ 1D) = 3\log L = \log V \tag{15}$$

Similarly, for the momentum of the particle we assume that the momentum (or the velocity) along the three axes x, y and z are independent. Hence, we write:

$$H_{max}(momentum\ in\ 3D) = 3H_{max}(momentum\ in\ 1D) \tag{16}$$

We combine the SMI of the locations and momenta of one particle in a box of volume V, taking into account the uncertainty principle. The result is:

$$H_{max}(location\ and\ momentum\ in\ 3D) = 3\log\left[\frac{L\sqrt{2\pi emk_BT}}{h}\right] \tag{17}$$

2.3.5. Step Four: The SMI of Locations and Momenta of N Independent and Indistinguishable Particles in a Box of Volume V

The next step is to proceed from one particle in a box to N independent particles in a box of volume V. Given the location (x, y, z), and the momentum (p_x, p_y, p_z) of one particle within the box, we say that we know the *micro-state* of the particle. If there are N particles in the box, and if their micro-states are independent, we can write the SMI of N such particles simply as N times the SMI of one particle, i.e.,

$$\text{SMI}(N \; independent \; particles) = N \times \text{SMI}(one \; particle) \tag{18}$$

This equation would have been correct if the micro-states of all the particles were independent. In reality, there are always correlations between the micro-states of all the particles; one is due to the *indistinguishability* between the particles, the second is due to *intermolecular interactions* between the particles.

For indistinguishable particles there are correlations between the events "one particle in i_1" "one particle in i_2" … "one particle in i_n", we can define the *mutual information* corresponding to this correlation. We write this as:

$$I(1; 2; \ldots; N) = \log N! \tag{19}$$

The SMI for N indistinguishable particles will then be:

$$H(N \; particles) = \sum_{i=1}^{N} H(one \; particle) - \log N!. \tag{20}$$

For the definition of the total mutual information, see Ben-Naim [7].

We can now write the final result for the SMI of N indistinguishable (but non-interacting) particles as:

$$H(N \; indistinguishable \; particles) = N \log \; V\left(\frac{2\pi m e k_B T}{h^2}\right)^{\frac{3}{2}} - \log N! \tag{21}$$

Using the Stirling approximation for $\log N!$ (note again that we use here the natural logarithm) in the form:

$$\log N! \approx N \log N - N \tag{22}$$

We have the final result for the SMI of N indistinguishable particles in a box of volume V and temperature T:

$$H(1, 2, \ldots N) = N \log \left[\frac{V}{N}\left(\frac{2\pi m k_B T}{h^2}\right)^{\frac{3}{2}}\right] + \frac{5}{2}N \tag{23}$$

This is a remarkable result. By multiplying the SMI of N particles in a box of volume V, at temperature T, by a constant factor ($k_B \log_e 2$), one gets the *thermodynamic entropy* of an ideal gas of simple particles. This equation was derived by Sackur [10] and by Tetrode [11] in 1912, by using the Boltzmann definition of entropy. Here, we have derived the entropy function of an ideal gas from the SMI. See also Ben-Naim [7].

One can convert this expression to the *entropy function* $S(E, V, N)$, by using the relationship between the total kinetic energy of the system, and the total kinetic energy of all the particles:

$$E = N\frac{mv^2}{2} = \frac{3}{2}Nk_B T \tag{24}$$

Hence, the explicit entropy function of an ideal gas is:

$$S(E, V, N) = Nk_B \ln \left[\frac{V}{N}\left(\frac{E}{N}\right)^{\frac{3}{2}}\right] + \frac{3}{2}k_B N\left[\frac{5}{3} + \ln\left(\frac{4\pi m}{3h^2}\right)\right] \tag{25}$$

We can use this equation as a *definition* of the entropy of an ideal gas of simple particles characterized by constant energy, volume and number of particles.

The next step in the definition of entropy is to add to the entropy of an ideal gas, the *mutual information* due to intermolecular interactions. We shall not need it here. The details are available in Reference [7].

3. Various Formulations of the Second Law

We shall distinguish between the *"thermodynamic"* formulations and the *probability* formulation of the Second Law. In my opinion the latter is a more general and a more useful formulation of the Second Law. There is a great amount of confusion regarding these two different formulation. For instance, one of the thermodynamic formulations is: when we remove a constraint, from a constrained equilibrium state of an isolated system, the entropy can only increase, it will never decrease. This "never" is in absolute sense. On the other hand, the probability formulation, as we expect, is of a statistical nature; the system will never return to its initial state. Here, "never" is not absolute but only "in practice."

In all of the examples of irreversible processes given in the literature, it is claimed that one *never* observes the reversal process spontaneously; the gas *never* condenses into a smaller region in space, two gases *never* un-mix spontaneously after being mixed, and heat *never* flows from a cold to a hot body, Figure 4.

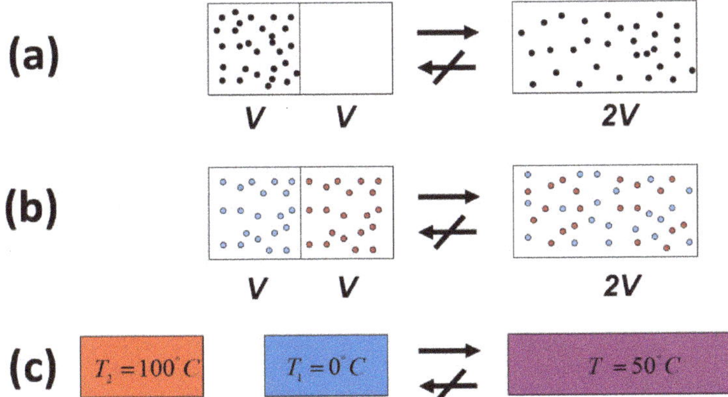

Figure 4. Three typical spontaneous irreversible processes occurring in isolated systems. (**a**) Expansion of an ideal gas; (**b**) Mixing of two ideal gases; (**c**) Heat transfer from a hot to a cold body.

Indeed, we *never* observe any of these processes occurring spontaneously in the reverse direction. For this reason, the processes shown in Figure 4 (as well as many others) are said to be *irreversible*. The idea of absolute *irreversibility* of these processes, was used to identify the direction of these process with the so-called Arrow of Time [4]. This idea is not only not true; it is also erroneously associated with the very definition of entropy. One should be careful with the use of the words "reversible" and "irreversible" in connection with the Second Law. Here, we point out two possible definitions of the term *irreversible*:

(1)   We *never* observe that the final state of any of the processes in Figure 4 returns back to the initial state (on the left-hand side of Figure 4) spontaneously.

(2)   We *never* observe the final state of any of the processes in Figure 4 going back to the initial state and *staying* in that state.

In case 1, the word *never* is used in "practice." The system can go from the final to the initial state. In this case, we can say that the initial state will be *visited*.

In case 2, the word *never* is used in an absolute sense. The system will never go back to its initial state and stay there!

In order to *stay* in one chamber, the partition should be replaced at its original position. Of course, this will *never* (in an absolute sense) occur spontaneously.

Notwithstanding the enormous success and the generality of the Second Law, Clausius made one further generalization of the Second Law:

The entropy of the universe always increases

This formulation is an unwarranted *over-generalization*. The reason is that the entropy of the universe cannot be defined!

3.1. Entropy-Formulations of the Second Law

In this section we shall discuss only one thermodynamic formulation of the Second Law for isolated systems. Other formulations are discussed in [4,6,13].

An isolated system, characterized by a fixed energy, volume and number of particles. For such a system the Second Law states [13]:

The entropy of an unconstrained isolated system (E, V, N), at equilibrium is larger than the entropy of any possible constrained equilibrium states of the same system.

An equivalent formulation of the Second Law is:

Removing any constraint from a constrained equilibrium state of an isolated system will result in an increase (or unchanged) of the entropy.

The maximum entropy is a maximum with respect to all *constrained equilibrium states*. This is very different from the maximum as a function of time, as it is often stated in the literature.

We quote here also the relationship between the change in entropy and the ratio of the probabilities:

$$\frac{\Pr(final)}{\Pr(initial)} = \exp\left[[S(final) - S(initial)]/k_{\mathrm{B}}\right]. \tag{26}$$

One should be very careful in interpreting this equality. The entropy change in this equation refers to change from state (a) to state (c), in Figure 5, both are equilibrium states. The probability ratio on the left hand side of the equation is for the states (b) and (c). The probability of state (a) as well as of state (c) is one! The reason is that both state (a) and state (c) are equilibrium states.

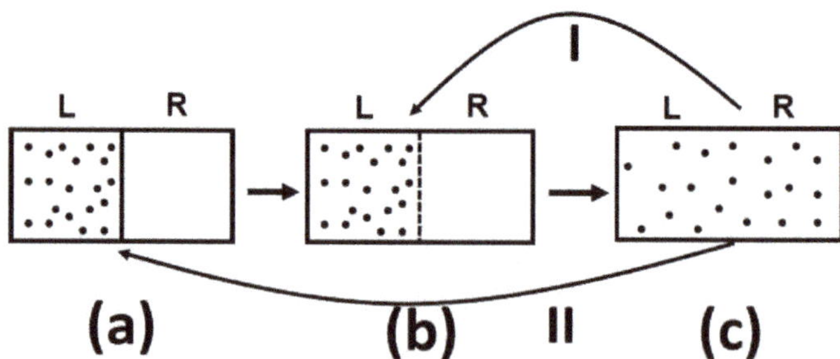

Figure 5. An expansion of an ideal gas. (**a**) The initial state; (**b**) The system at the moment after removal of the partition and (**c**) the final equilibrium state.

3.1.1. Probability Formulations of the Second Law for Isolated Systems

We present here these three equations which relate the difference in some thermodynamic quantity, with the ratio of probabilities:

$$\frac{\Pr(final)}{\Pr(initial)} = \exp\Big[[S(final) - S(initial)]/^{k_B}\Big]$$
$$\frac{\Pr(final)}{\Pr(initial)} = \exp\Big[-[A(final) - A(initial)]/^{k_B T}\Big] \tag{27}$$
$$\frac{\Pr(final)}{\Pr(initial)} = \exp\Big[-[G(final) - G(initial)]/^{k_B T}\Big]$$

The first is valid for an isolated system, the second for a system at constant temperature, and the third for a system at constant temperature and pressure.

It is clear from Equations (27) that the probability formulation of the Second Law, which will state below, is far more general than any of the thermodynamic formulations in terms of either entropy, Helmholtz energy or Gibbs energy.

The origin of the probability formulation of the Second law can be traced back to Boltzmann:

" ... *the system ... when left to itself, it rapidly proceeds to the disordered, most probable state.*"

Note that here Boltzmann uses the term disorder to describe what happens "when (the system) is left to itself, it rapidly proceeds to disordered most probable state.

3.1.2. The Probability Formulation of the Second Law

Why do we see the process going in one direction? Clearly, the random motion of the particles cannot determine a unique direction. The fact that we observe such a one-way, or one-directional processes led many to associate the so-called Arrow of Time with the Second Law, more specifically with the "tendency of entropy to increase."

It we ask: "What is the cause of the one-way processes?" or "What drives the processes in one direction?" The answer is probabilistic; the processes are not absolutely irreversible. All these processes are irreversible only *in practice*, or equivalently with *high probability*.

Note that *before* we removed the constraint, the probability of finding the initial state is *one*. This is an equilibrium state, and all the particles are, by definition, of the initial state. However, after the removal of the constrain, we have for N of the order of 10^{23}:

$$\frac{\Pr(final\ configuration)}{\Pr(initial\ configuration)} \approx infinity \tag{28}$$

This is essentially the probability formulation of the Second Law for this particular experiment. This law states that starting with an equilibrium initial state, and removing the constraint (the partition), the system will evolve to a new equilibrium configuration which has a probability overwhelmingly larger than the initial configuration.

Let us repeat the probability formulation of the Second Law for this particular example. We start with an initial *constrained equilibrium state*. We remove the constraint, and the system's configuration will evolve with probability (nearly) one, to a new equilibrium state, and we shall *never* observe reversal to the initial state. "Never" here, means never in our lifetime, or in the lifetime of the universe.

3.1.3. Some Concluding Remarks

At the time when Boltzmann proclaimed the probabilistic approach to the Second Law, it seemed as if this law was somewhat *weaker* than the other laws of physics. All physical laws are absolute and no exceptions are allowed. The Second Law, as formulated by Clausius was also absolute. On the other hand, Boltzmann's formulation was not absolute–exceptions were allowed. Much later came the realization that the admitted non-absoluteness of Boltzmann's formulations of the Second Law, was in fact more *absolute* than the absoluteness of the macroscopic formulation of the Second Law, as well as of any other law of physics for that matter. It should be noted that although Boltzmann was right in claiming that the system tends to the most probable state, he erred when he claimed that the system tends to a disordered state. More details on this in Ben-Naim [4].

What about the violations of the Second Law? Here, one should be prepared for a shocking statement.

Most writers who write about the Second Law, and who recognize its statistical nature, would say that since the system can return to its initial state, the violation of the Second Law is possible. They might add that such "violations" could occur for small N, but it will *never* occur for large N. Thus, one concludes that the Second Law, because of its statistical nature can be violated.

I say, No! The Second Law, in any formulation will *never* be violated, never in the absolute sense, not for small N, and not for very large N!

This sounds "contradictory" to the very statistical nature of the probability formulation of the Second Law. However, it is not. The reason is that the occurrence of an extremely improbable event is not a violation of the law which states that such events have very low probability.

Thus, the system can return to the initial state, in this sense the Second Law is indeed a statistical Law. However, whenever that (very rare) event occurs it is not a violation of the Second Law.

Note also that the Entropy formulation of the Second Law is not statistical. Entropy will not decrease in a spontaneous process in an isolated system. Thus, we can conclude the Second Law, in either formulation, can never be violated, *never* in the absolute sense!

4. Interpretation and Misinterpretations of Entropy.

In this section we present only a few misinterpretations of entropy. More details may be found in [4].

4.1. The Order-Disorder Interpretation

The oldest and the most common interpretation of entropy, sometimes also used as a "definition" of entropy, is *disorder*. It is unfortunate that this interpretation has survived until these days in spite of being falsified in several books and articles, Ben-Naim [4].

It is not clear who was the first to equate entropy with disorder. The oldest association of disorder with the Second Law is probably in Boltzmann's writings. Here are some quotations [14,15]:

" ... *the initial state of the system ... must be distinguished by a special property (ordered or improbable) ...* "

" ... *this system takes in the course of time states ... which one calls disordered.*"

"*Since by far most of the states of the system are disordered, one calls the latter the probable states.*"

" ... *the system ... when left to itself, it rapidly proceeds to the disordered, most probable state.*"

In a delightful book *Circularity*, by Aharoni [16], on page 41 we find:

" ... *disorder (measured by the parameter called entropy) increases with time*"

Interestingly, this statement appears in a book on circularity. Here disorder is "*measured by entropy*", and entropy is a measure of disorder. A perfect example of circularity. I am bringing this quotation from Aharoni's excellent book, not as a criticism of the book, but as an example of an excellent writer who simply took a wrong idea from the literature and expanded it: On page 42 we find:

"*Nobody know for sure why the world is going from order to disorder, but this is so*"

This is a typical statement, where one takes for granted the "truth" that the world goes from order to disorder, then expressed only the puzzlement about not knowing why this "fact" is so. Of course no one is sure about that "fact" as no one is sure why the world is going from being beautiful to uglier. The answer is very simple: I am sure that the world is not going from order to disorder. Thus the very assumption made in this quoted statement is definitely not true!

4.2. The Association of Entropy with Spreading/Dispersion/Sharing

The second, most popular descriptor of entropy as "spreading" was probably suggested by Guggenheim [17]. Guggenheim started with the Boltzmann definition of entropy in the form: $S(E) = k \log \Omega(E)$, where k is a constant and: "$\Omega(E)$ *denotes the number of accessible independent quantum states of energy E for a closed system.*"

> *"To the question what in one word does entropy really mean, the author would have no hesitation in replying 'Accessibility' or 'Spread.' When this picture of entropy is adopted, all mystery concerning the increasing property of entropy vanishes."*

Authors who advocate the "spreading" interpretation of entropy would also say that when the energy spread is larger, the entropy change should be larger too. Therefore, the change in the entropy in the expansion of one mole of an ideal gas, from volume V to $2V$, in either an isolated system, or in an isothermal process is $\Delta S = R \ln \frac{2V}{V} = R \ln 2$ where n is the number of moles, R is the gas constant. You can repeat the same process at different temperatures, as long as you have an ideal gas, the change in entropy in this process is the same $nR \ln 2$, independently of the temperature (as well as on the energy of the gas). Thus, as long as we have an ideal gas the change of entropy in the expansion of one mole of gas from V to $2V$ is *independent* of temperature. It is always $R\ln2$.

4.3. Entropy as Information; Known and Unknown, Visible and Invisible and Ignorance

The earliest explicit association between information and entropy is probably due to Lewis [18]:

> *"Gain in entropy always means loss of information and nothing more."*

It is clear, therefore that Lewis did not use the term "information" is the sense of SMI. The misinterpretation of the information-theoretical entropy as a subjective information is quite common. Here is a paragraph from Atkins' preface from the book [2] "*The Second Law*".

> *"I have deliberately omitted reference to the relation between information theory and entropy. There is the danger, it seems to me, of giving the impression that entropy requires the existence of some cognizant entity capable of possessing 'information' or of being to some degree 'ignorant.' It is then only a small step to the presumption that entropy is all in the mind, and hence is an aspect of the observer."*

Atkins' comment and his rejection of the informational interpretation of entropy on the grounds that this "relation" might lead to the "presumption that entropy is all in the mind" is ironic. Instead, he uses the terms "disorder" and "disorganized," etc., which are concepts that are far more "in the mind."

4.4. Entropy as a Measure of Probability

To begin with, let me say the following: ***Entropy is not probability!***
Numerous authors would start by writing Boltzmann's equation for entropy as:

$$S = k \ln P \tag{29}$$

where P is supposed to stand for probability. Of course, P cannot be probability; Probability, is a number smaller than 1, hence, the entropy would become a negative number.

Brillouin was perhaps the first who wrote Boltzmann's equation in this form. On page 119 of Brillouin's book [19] we find:

> *Entropy is shown to be related to probability. A closed isolated system may have been created artificially with very improbable structure ...*

Next comes the most absurd, I would also say, *shameful* statement which should not be made by anyone who has any knowledge of probability. On page 120 we find:

> *The probability has a natural tendency to increase, and so does entropy. The exact relation is given by the famous Boltzmann-Planck formula:*

$$S = k \ln P$$

As anyone who knows elementary probability theory, probability does not have (a natural or unnatural) tendency to increase (or to decrease)! Entropy does not have a tendency to increase! And there is no such a relationship between entropy and probability! This is plainly a shameful statement, nothing more.

4.5. Entropy as a Measure of Irreversibility

The Second Law is usually discussed in connection with the apparent irreversibility of natural processes. Although the terms reversible and irreversible have many interpretations it is very common to say the that $\Delta S > 0$ in the Clausius formulation is a measure of irreversibility.

In a recent book by Rovelli [20] we find on page 25 where the author introduces the entropy as:

> *Clausius introduces a quantity that measures this irreversible progress of heat in only one direction and, since he was a cultivated German, he gives it a name taken from ancient Greek–entropy.*

It is such a shame that a theoretical physicist would write such a sentence in a book published in 2018! To see the fallacy of such a statement, just have a look at Figure 5.

I tell you that in system (a), there are N particles at a given volume V and temperature T. In (c) the same number of particle are in volume $2V$ and at the same temperature T. I will also tell you that the system (c) has higher entropy that system (a). Can you tell which system is more or less "irreversible?"

In a glossary of Lemon's book [21], we find:

> *"Entropy: A measure of the irreversibility of the thermodynamic evolution of an isolated system."*

This is not true! First, because irreversibility means several things. Second, because irreversibility is used to describe a process, whereas entropy is a quantity assigned to a *state* of a system. Third, because entropy is a measurable quantity, but "irreversibility," is not a measurable quantity.

Neither the value of the entropy of a system, nor the change in entropy can be a measure of the irreversibility–certainly not the "irreversibility of the thermodynamic evolution of an isolated system."

Entropy, in itself is a state function. This means that for any well-defined thermodynamic system, say (T, P, N), the entropy is also determined. Specifically, for an isolated system defined by the constant variables E, V, N, the entropy is fixed. The value of S has nothing to do with the "evolution" of the system, nor with any process that the system is going through.

In the text of Lemon's book, we find a different "definition." On page 9, the author states:

> *" … the entropy difference between two states of an isolated system quantifies the irreversibility of a process connecting those two states"*

Indeed, the entropy *difference* is by definition positive for a spontaneous process in an isolated system. However, entropy difference between to states of isolated system, is a quantity fixed by the two states. It does not depend on how we proceed from one state to another; *Reversibility* or *irreversibility* of a process is not quantifiable.

4.6. Entropy as Equilibriumness

Here is another amusing idea about the "meaning" of entropy. This "original," and as far as I know, a unique "invention" appears in Seife's book *"Decoding the Universe"* [22]. On page 32 of Seife's book we find the obscure sentence:

> *"The equilibrium of the universe increases, despite your best effort."*

Can anyone tell me what does it means that the "equilibrium" of anything (let alone the universe) increases (with or without our efforts)? In the next paragraph the author introduces another meaningless term, which is supposed to "explain" the previous obscure sentence:

"What if you don't use an engine? What if you can turn a crank by hand? Well, in actuality your muscles are acting as an engine, too. They are exploiting the chemical energy stored in molecules in your bloodstream, breaking them apart, and releasing the energy into the environment in the form of work. This increases the "equilibriumness" of the universe just as severely as a heat engine does."

As I have noted in my book [4], such statements are a result of a profound misunderstanding of the Second Law. There are many other interpretations of entropy, these are discussed in [4].

5. Misuses and Misapplications of Entropy and the Second Law

In this section we shall very briefly review three main misuses of entropy and the Second Law. More details in Ben-Naim [4].

5.1. The Association of Entropy with Time

The origin of the association of entropy with "Time's Arrow" can be traced to Clausius' famous statement of the Second Law:

"The entropy of the universe always increases."

The statement *"entropy always increases"*, means implicitly that *"entropy always increases with time"*. However, it is Eddington [23] who is credited for the *explicit* association of "The law that entropy always increases" with "Time's Arrow", which expresses this "one-way property of time." See below.

There are two very well-known quotations from Eddington's book, *"The Nature of the Physical World"* [23]. One concerns the role of entropy and the Second Law, and the second, introduces the idea of "time's arrow." In the first quotation Eddington reiterates the unfounded idea that "entropy always increases." Although it is not explicitly stated, the second alludes to the connection between the Second Law and the Arrow of Time. This is clear from the association of the *"random element in the state of the world"* with the *"arrow pointing towards the future"*.

There are many other statements in Eddington's book which are unfounded and misleading. For instance; the claim that entropy is a *subjective* quantity, the concepts of *"entropy-clock"*, and *"entropy-gradient"*. Reading through the entire book by Eddington, you will not find a single correct statement on the thermodynamic entropy!

5.2. Boltzmann's H-Theorem and the Seeds of an Enormous Misconception about Entropy

In 1877 Boltzmann proved a truly remarkable theorem, known as the H-theorem. He defined a function $H(t)$, and proved that it decreases with time and reaches a minimum at equilibrium. In the following shall we briefly present the H-theorem, the main criticism, and Boltzmann's answer. We will point out where Boltzmann went wrong and why the function $-H(t)$ is not entropy, and why the H-theorem does not represent the Second Law.

Boltzmann defined a function $H(t)$ as:

$$H(t) = \int f(v,t) \log[f(v,t)]dv \qquad (30)$$

Then Boltzmann made a few assumptions. Details of the assumptions and the proof of the theorem can be found in many textbooks. Basically, Boltzmann proved that:

$$\frac{dH(t)}{dt} \leq 0 \qquad (31)$$

and at equilibrium, i.e., $t \to \infty$:

$$\frac{dH(t)}{dt} = 0 \tag{32}$$

Boltzmann believed that the behavior of the function $-H(t)$ is the same as that of the entropy, i.e., the entropy always increases with time, and at equilibrium, it reaches a maximum, thereafter it does not change with time.

This theorem drew great amount of criticism. We shall not discuss these here. It is sufficing to say that Boltzmann correctly answered all the criticism, see [4,5]. Notwithstanding Boltzmann's correct answers to his critics, Boltzmann and his critics made an enduring mistake in the interpretation of the H-function, a lingering mistake that has hounded us ever since. This is the very identification of the H-Theorem with the behavior of the entropy. It is clear from the very definition of the function $H(t)$, that $-H(t)$ is a SMI. If one identifies the SMI with entropy, then we go back to Boltzmann's identification of the function $-H(t)$ with entropy.

To obtain the entropy one must first define the $-H(t)$ function based on the distribution of both the locations and momentum, i.e.,:

$$-H(t) = -\int f(R,p,t)\log f(R,p,t)dRdp \tag{33}$$

To obtain the entropy we must take the maximum of $-H(t)$ over all possible distributions $f(R,p,t)$:

$$Entropy = \max_{over\ all\ fs}\left[-H(t)\right] \tag{34}$$

We believe that once the system attains an equilibrium, the $-H(t)$ attains its maximum value, i.e., we identify the *maximum* over all possible distributions with the maximum of SMI in the limit $t \to \infty$, i.e.,

$$Entropy = \lim_{t\to\infty}\left[-H(t)\right] = Max\ SMI\ (at\ equilibrium) \tag{35}$$

At this limit we obtain the entropy (up to a multiplicative constant), which is clearly not a function of time! Thus, once it is understood that the function $-H(t)$ is an SMI and not entropy, it becomes clear that the criticism of Boltzmann's H-Theorem is addressed to the evolution of the SMI, and not of the entropy.

In responding the editor's comment, I would like to add that I am well aware that there exists an extensive literature on non-equilibrium thermodynamics. My conclusion in this article is based on examination of the *definitions* of entropy and the *formulation* of the Second Law. In a separate article submitted to Entropy, I have shown that in all the literature on non-equilibrium thermodynamics all the authors introduce the assumption of "local equilibrium" without giving any justification. I believe that this fact casts some serious doubts on the applicability of thermodynamics to systems which are *very far* from equilibrium.

5.3. Can Entropy Be Defined for, and the Second Law Applied to Living Systems?

This question has been discussed by numerous physicists, in particular by Schrödinger, Monod, Prigogine, Penrose and many others [4].

One cannot avoid starting with the most famous book written by Schrödinger [24]: What is life?

Chapter 6 titled "Order, disorder and entropy." He starts with the common statement of the Second Law in terms of the "order" and "disorder"

"It has been explained in Chapter 1 that the laws of physics, as we know them, are statistical laws. They have a lot to do with the natural tendency of things to go over into disorder."

His main claim is that "living matter evades the decay to equilibrium." Then he asks:

"How does the living organism avoid decay? The obvious answer is: By eating, drinking, breathing and (in the case of plants) assimilating. The technical term is metabolism."

I believe the highlight of the book is reached on page 76:

"What then is that precious something contained in our food which keeps us from death? That is easily answered. Every process, event, happening–call it what you will; in a word, everything that is going on in Nature means an increase of the entropy of the part of the world where it is going on. Thus, a living organism continually increases its entropy–or, as you may say, produces positive entropy–and thus tends to approach the dangerous state of maximum entropy, which is death. It can only keep aloof from it, i.e., alive, by continually drawing from its environment negative entropy–which is something very positive as we shall immediately see. What an organism feeds upon is negative entropy."

Such statements, in my opinion are nothing but pure nonsense. Entropy, by definition, is a positive quantity. There is no negative entropy!

There are many statements in popular science which relate biology to ordering, and ordering as decrease in entropy. Here is an example:

Atkins [2], in his introduction to the book writes:

"In Chapter 8 we also saw how the Second Law accounts for the emergence of the intricately ordered forms characteristic of life."

Of course, this is an unfulfilled promise. No one has ever shown that the Second accounts for anything associated with life. Brillouin [19], further developed the idea of "feeding on the negative entropy" and claimed that:

"If living organism needs food, it is only for the negentropy it can get from it, and which is needed to make up for the losses due to mechanical work done, or simple degradation processes in living systems. Energy contained in food does not really matter: Since energy is conserved and never gets lost, but negentropy is the important factor."

In a recent book by Rovelli [20], the nonsensical idea that "entropy is more important than energy, is elevates to highest peak. You will find there a statement written in all capital letters:

IT IS ENTROPY, NOT ENERGY THAT DRIVES THE WORLD.

5.4. Entropy and Evolution

In an article entitled: *"Entropy and Evolution"*, Styer [25] begins with a question, "Does the Second Law of thermodynamics prohibit biological evolution?" Then he continues to show "quantitatively" that there is no conflict between evolution and the Second Law. Here is how he calculates the "entropy required for evolution." Suppose that, due to evolution, each individual organism is 1000 times "more improbable" than the corresponding individual was 100 years ago. In other words, if Ω_i is the number of microstates consistent with the specification of an organism 100 years ago, and Ω_f is the number of microstates consistent with the specification of today's "improved and less probable" organism, then $\Omega_f = 10^{-3}\Omega_i$." From these two numbers he estimates the change in entropy per one evolving organism, then he estimates the change in entropy of the entire biosphere due to evolution. His conclusion:

"The entropy of the earth's biosphere is indeed decreasing by a tiny amount due to evolution, and the entropy of the cosmic microwave background is increasing by an even greater amount to compensate for that decrease."

In my opinion this *quantitative* argument is superfluous. In fact, it weakens the *qualitative* arguments I have given above. No one knows how to calculate the "number of states" (Ω_i and Ω_f) of any living organism. No one knows *what the states* of a living organism are, let alone *count* them. Therefore, the estimated change in entropy due to evolution is meaningless.

5.5. Application of Entropy and the Second Law to the Entire Universe

We already mentioned Clausius' *over-generalizing* the Second Law. His well-known and well quoted statement:

"The entropy of the universe always increases."

I do not know how Clausius arrived at this formulation of the Second Law. Clausius did not, and in fact could not understand the meaning of entropy. One can also safely cay that Clausius' formulation of the Second Law for the entire universe, is not justified and in fact cannot be justified.

In many books, especially popular science books, one finds a statement of the Second Law as:

Entropy always increases.

This statement is sometimes used either a statement of the Second Law or a short version of Clausius' statement:

Entropy of the universe always increases.

The first statement is obviously meaningless; entropy, in itself does not have a numerical value. Therefore, it is meaningless to say that it increases or decreases.

The second statement seems more meaningful because it specifies the "system" for which the entropy is said to increase. Unfortunately, this is also meaningless since the entropy of the universe is not definable.

On page 299 of Carroll's book [26], after talking too much about entropy, the author poses a simple question:

"So what is its entropy?"

"Its" refers to the "universe." Then the author provides some numbers. First, he says that the early universe was "just a box of hot gas," and a "box of hot gas is something whose entropy we know how to calculate." No, we do not know! We know how to calculate the entropy of *ideal* gas, i.e., non-interacting particles, i.e., very dilute gas, not extremely dense gas that was supposed to be in the early universe. Then he provides some numbers:

$$S_{early} \approx 10^{83}$$

The author explains that the "$\approx$" sign mean: *"approximately equal to,"* and that he wants to emphasize that this is only a rough estimate, not a rigorous calculation:

This number comes from simply treating the contents of the universe as a conventional gas in thermal equilibrium."

This is *not* an *approximate* number. On the contrary, this is *exactly* a *meaningless number*. The content of the early universe cannot be treated as "conventional gas in thermal equilibrium." Then, we are given another number of the entropy:

$$S_{today} \approx 10^{121}$$

and, yet another number:

$$S_{max} \approx 10^{120}$$

All these numbers are meaningless. After finishing with the numbers, the absurdities are taken into new heights. On page 301, we find:

"The conclusion is perfectly clear: The state of the early universe was not chosen randomly among all possible states. Everyone in the world who has thought about the problem agrees with that."

Everyone? I certainly do not wish to be included as one who has thought about this problem. I never thought about the meaningless entropy of the early universe.

On page 311, we find another "high entropy" absurd question and answer:

"Why don't we live in empty space?"

I honestly could not believe that a scientist can pose such a silly and laughable question. Then on page 43 of the book we find:

When it comes to the past, however, we have at our disposal both our knowledge of the current macroscopic state of the universe, plus the fact that the early universe began in a low-entropy state. That one extra bit of information, known simply as the "Past Hypothesis," gives us enormous leverage when it comes to reconstructing the past from the present.

Thus, Carroll, not only assigned numbers to the entropy of the universe at present, and not only estimated the entropy of the universe at the speculative event referred to as the Big Bang, but he also claimed that this low-entropy "fact" can explain many things such as "why we remember the past but not the Future … "

All these senseless claims could have been averted has Carroll, as well as many others recognize that it is meaningless to talk about the entropy of the universe. It is a fortiori meaningless to talk about the entropy of the universe at some distant point in time.

6. Conclusions

Starting with the very definitions of entropy it is clear that entropy is a well-defined *state function*. As such it is well-defined for any system at equilibrium. It is also clear that entropy is not a function of time and it is not related to Time's Arrow. It is also clear that entropy cannot be used for any living system or to the entire universe. As one of the reviewers suggested, the concept of entropy has expanded from heat machines to realms in which it cannot even be defined. This is precisely why I quoted Einstein on thermodynamics. It is true that thermodynamics will not be *overthrown*, provided it is applied *"within the framework of applicability"*.

Funding: This research received no external funding.

Conflicts of Interest: The author declares no conflict of interest.

References

1. Shermer, M. *Heavens on Earth: The Scientific Search for the Afterlife, Immortality, and Utopia*; Henry Holt and Co.: New York, NY, USA, 2018.
2. Atkins, P. The Second Law. In *Scientific American Books*; W.H. Freeman and Co.: New York, NY, USA, 1984.
3. Atkins, P. *Four Laws That Drive the Universe*; Oxford University Press: Oxford, UK, 2007.
4. Ben-Naim, A. *The Greatest Blunder Ever in the History of Science, Involving: Entropy, Time Life and the Universe*; 2020; in preparation.
5. Ben-Naim, A. *Time's Arrow (?) The Timeless Nature of Entropy and the Second Law of Thermodynamics*; Lulu Publishing Services: Morrisville, NC, USA, 2018.
6. Ben-Naim, A.; Casadei, D. *Modern Thermodynamics*; World Scientific Publishing: Singapore, 2017.
7. Ben-Naim, A. *Information Theory, Part I: An Introduction to the Fundamental Concepts*; World Scientific Publishing: Singapore, 2017.
8. Ben-Naim, A. *Entropy the Truth the Whole Truth and Nothing but the Truth*; World Scientific Publishing: Singapore, 2016.
9. Ben-Naim, A. *The Four Laws That Do not Drive the Universe*; World Scientific Publishing: Singapore, 2017.
10. Sackur, O. *Annalen der Physik*; WILEY-VCH Verlag GmbH &, Co.: Weinheim, Germany, 1911; Volume 36, p. 958.
11. Tetrode, H. *Annalen der Physik*; WILEY-VCH Verlag GmbH &, Co.: Weinheim, Germany, 1912; Volume 38, p. 434.

12. Shannon, C.E. A Mathematical Theory of Communication. *Bell System Tech. J.* **1948**, *27*, 379–423. [CrossRef]
13. Callen, H.B. *Thermodynamics and an Introduction to Thermostatics*, 2nd ed.; Wiley: New York, NY, USA, 1985.
14. Brush, S.G. *The Kind of Motion We Call Heat. A History of The Kinetic Theory of Gases in the 19th Century, Book 2: Statistical Physics and Irreversible Processes*; North-Holland Publishing Company: Amsterdam, The Netherlands, 1976.
15. Brush, S.G. *Statistical Physics and the Atomic Theory of Matter, from Boyle and Newton to Landau and Onsager*; Princeton University Press: Princeton, NJ, USA, 1983.
16. Aharoni, R. *Circularity; A Common Secret to Paradoxes, Scientific Revolutions and Humor*; World Scientific Publishing: Singapore, 2016.
17. Guggenheim, E.A. Statistical Basis of Thermodynamics. *Research* **1949**, *2*, 450.
18. Lewis, G.N. The Symmetry of Time in Physics. *Science* **1930**, *71*, 569. [CrossRef] [PubMed]
19. Brillouin, L. *Science and Information Theory*; Academy Press: New York, NY, USA, 1962.
20. Rovelli, C. *The Order of Time*; Riverhead Books: New York, NY, USA, 2018.
21. Lemons, D.S. *A Student's Guide to Entropy*; Cambridge University Press: Cambridge, UK, 2013.
22. Seife, C. *Decoding the Universe. How the Science of Information is Explaining Everything in the Cosmos, From our Brains to Black Holes*; Penguin Book: East Rutherford, NJ, USA, 2006.
23. Eddington, A. *The Nature of the Physical World*; Cambridge University Press: Cambridge, UK, 1928.
24. Schrödinger, E. *What Is Life? The Physical Aspect of the Living Cell*; Cambridge University Press: Cambridge, UK, 1944.
25. Styer, D.F. Entropy and Evolution. *Am. J. Phys.* **2008**, *76*, 1031. [CrossRef]
26. Carroll, S. *From Eternity to Here, The Quest for the Ultimate Theory of Time*; Plume: New York, NY, USA, 2010.

Article

Changes of Conformation in Albumin with Temperature by Molecular Dynamics Simulations

Piotr Weber [1,†], **Piotr Bełdowski** [2,*,†], **Krzysztof Domino** [3,†], **Damian Ledziński** [4,†] and **Adam Gadomski** [2,†]

1. Atomic and Optical Physics Division, Department of Atomic, Molecular and Optical Physics, Faculty of Applied Physics and Mathematics, Gdańsk University of Technology, ul. G. Narutowicza 11/12, 80-233 Gdańsk, Poland; piotr.weber@pg.edu.pl
2. Institute of Mathematics and Physics, UTP University of Science and Technology, Kaliskiego 7, 85-796 Bydgoszcz, Poland; agad@utp.edu.pl
3. Institute of Theoretical and Applied Informatics, Polish Academy of Sciences, Bałtycka 5, 44-100 Gliwice, Poland; kdomino@iitis.pl
4. Faculty of Telecommunications, Computer Science and Technology, UTP University of Science and Technology, 85-796 Bydgoszcz, Poland; damian.ledzinski@utp.edu.pl
* Correspondence: piobel000@utp.edu.pl
† These authors contributed equally to this work.

Received: 16 January 2020; Accepted: 29 March 2020; Published: 1 April 2020

Abstract: This work presents the analysis of the conformation of albumin in the temperature range of $300K$–$312K$, i.e., in the physiological range. Using molecular dynamics simulations, we calculate values of the backbone and dihedral angles for this molecule. We analyze the global dynamic properties of albumin treated as a chain. In this range of temperature, we study parameters of the molecule and the conformational entropy derived from two angles that reflect global dynamics in the conformational space. A thorough rationalization, based on the scaling theory, for the subdiffusion Flory–De Gennes type exponent of 0.4 unfolds in conjunction with picking up the most appreciable fluctuations of the corresponding statistical-test parameter. These fluctuations coincide adequately with entropy fluctuations, namely the oscillations out of thermodynamic equilibrium. Using Fisher's test, we investigate the conformational entropy over time and suggest its oscillatory properties in the corresponding time domain. Using the Kruscal–Wallis test, we also analyze differences between the root mean square displacement of a molecule at various temperatures. Here we show that its values in the range of $306K$–$309K$ are different than in another temperature. Using the Kullback–Leibler theory, we investigate differences between the distribution of the root mean square displacement for each temperature and time window.

Keywords: Flory–De Gennes exponent; conformation of protein; albumin; non-gaussian chain; non-isothermal characteristics; Fisher's test; Kullback–Leibler divergence

---

1. Introduction

The dynamics of proteins in solution can be modeled by a complex physical system [1], yielding many interesting effects such as the competition and cooperation between elasticity and phenomena caused by swelling. Biopolymers, such as proteins, consist of monomers (amino acids) that are connected linearly. Thus, a chain seems to be the most natural model of such polymers. However, depending on the properties of the polymer molecule, modeling it as a chain may give more or less accurate outcomes [2]. The most straightforward approach to the polymer is the freely jointed chain, where each monomer moves independently. The excluded volume chain effect, which prevents a polymer's segments from overlapping, is one of the reasons why the real polymer chain differs from an ideal chain [3]. The diffusive dynamics of biopolymers are discussed further in [4] and the references therein, and the subdiffusive dynamics are discussed in [5,6]. For a given combination of polymer and solvent, the excluded volume varies with the temperature and can be zero at a specific temperature, named the Flory temperature [7]. In other words, at the Flory temperature, the polymer chain becomes nearly ideal [8]. However, this parameter is global, and a more detailed description of the dynamics is required to perform additional detailed analyses. Our work concentrates on an analysis of albumin (shown in Figure 1) [9]. Albumin plays several crucial roles as the main blood plasma protein (60% of all proteins): controlling pH, inducing oncotic pressure, and transporting fluid. This protein (of nonlinear viscoelastic, with characteristics of rheopexy [10]) plays an essential role in the process of articular cartilage lubrication [11] by synovial fluid (SF). SF is a complex fluid of crucial relevance for lowering the coefficients of friction in articulating joints, thus facilitating their lubrication [12]. From a biological point of view, this fluid is a mixture of water, proteins, lipids, and other biopolymers [13]. As an active component of lubricating mechanisms, albumin is affected by friction-induced increases of temperature in synovial fluid. The temperature inside the articular cartilage is in the range of $306K$–$312K$ [14], and changes occurring inside the fluid can affect the outcome of albumin's impact on lubrication [10]. Albumin can help to regulate temperature during fever. Therefore, a similar mechanism could take place during joint friction, increasing SF lubricating properties as an efficient heat removal. We analyze this range of temperature: from $300K$–$312K$.

Let us move to the technical introduction of albumin dynamics. Consider a chain that was formed by $N+1$ monomers, assuming the following positions, designated by $\{\vec{r}_0, \vec{r}_1, \ldots, \vec{r}_N\}$. We denote by $\vec{\tau}_j = \vec{r}_j - \vec{r}_{j-1}$ a vector of a distance between two neighboring monomers. While considering the albumin, we have $N+1 = 579$ amino acids. For the length of the distance vector $\vec{\tau}_j$, we take the distance between two α-carbons in two consecutive amino acids, hence we have $|\vec{\tau}_j| = \tau = 3.8$ [Å] [15]. Using such a notation, we can define the end-to-end vector $\vec{R}$ by the following formula:

$$\vec{R} = \vec{r}_N - \vec{r}_0 = \sum_{j=1}^{N} \vec{\tau}_j. \tag{1}$$

In the case of an ideal chain, the distance vectors (for all Ns) are completely independent of each other. This property is expressed by a lack of correlation between any two different bonds, that is [16]:

$$\langle \vec{\tau}_i \cdot \vec{\tau}_j \rangle = \tau^2 \delta_{ij}, \tag{2}$$

where δ_{ij} is the Kronecker delta. Equations (1) and (2) imply that a dot self-product of vector $\vec{R}$ reads:

$$\langle \vec{R} \cdot \vec{R} \rangle = \langle R^2 \rangle = \sum_{i=1}^{N} \sum_{j=1}^{N} \langle \vec{\tau}_i \cdot \vec{\tau}_j \rangle = N\tau^2, \tag{3}$$

where R is the length of the vector $\vec{R}$. The value of R is measured by calculating the root mean square end-to-end distance of a polymer. The R'th value is assumed to depend on the polymer's length N in the following manner [17]:

$$R \sim \tau N^{\mu}, \tag{4}$$

where μ is called the size exponent. If a chain is freely joined, it appears that $\mu = 1/2$. Relation (4) suggests we consider the dynamics of the albumin protein in some range of temperature and look for dependence between the temperature and exponent μ. Applying the natural-logarithm operation to (4) yields:

$$\mu \simeq \frac{\log R - \log \tau}{\log N}. \tag{5}$$

In Equation (4) we use a $\sim$ symbol to emphasize that we use $\langle R^2 \rangle$ to obtain R. In the case discussed here, both parameters τ and N (the number of amino acids) are constant. Importantly, using molecular dynamics simulations, one can obtain the root mean square end-to-end distance for various temperatures. The idea of the chain points to its description by backbone angles and dihedral angles. Backbone angles are angles between three α-carbons occurring after each other and signed by ϕ. The second angle, designated by ψ, is a dihedral angle [18]. Using this parametrization, we can obtain an experimental distribution of probabilities. Such distributions have features that may display some time-dependent changes.

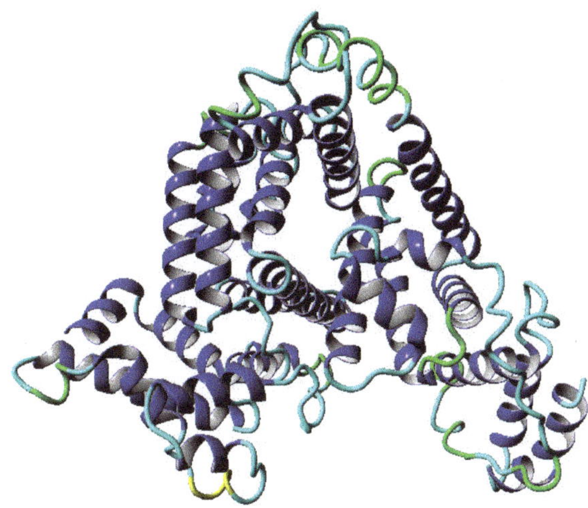

Figure 1. Structure of the albumin protein in a ribbon-like form. The colors on the molecular surface indicate the secondary structure: blue—α-helix, green—turn, yellow—3–10 helix, cyan—coil.

These time-dependent changes may have an utterly random course, but sometimes periodic behavior can be seen. This property can be analyzed using a particular statistical test [19]. There is an assumption that the time series sampling is even, and the test treats it as a realization of the stochastic model evolution, which consists of specific terms of the form:

$$y_n = \beta \cos (n\omega + \alpha) + \epsilon_n, \tag{6}$$

where $\beta \geq 0$, $n = 0, 1, 2, \ldots, N$, $0 < \omega < \pi$, $-\pi < \alpha \leq \pi$ is uniformly distributed in $(-\pi; \pi]$ phase factor and ϵ_n is an independent, identically distributed random noise sequence. The null hypothesis is that amplitude β is equal zero: (H_0: $\beta = 0$) against the alternative hypothesis that H_1: $\beta \neq 0$.

A periodogram of a time series $(y_0, y_1, \ldots, y_N)$ is used to test the above mentioned hypothesis and it is defined by the following formula:

$$I(\omega) = \frac{1}{N} \left| \sum_{j=0}^{N} \exp(-i\omega j) y_j \right|^2. \tag{7}$$

where in general, $\omega \in [0, \pi]$ and i is the imaginary unit. For a discrete time series, a discrete value of ω from $[0, \pi]$ is being used:

$$\omega_k = \frac{2\pi k}{N}, \qquad k = 0, 1, \ldots, N, \tag{8}$$

which are known as Fourier frequencies.

The periodogram (for a single time series) can contain a peak. The statistical test should give us information about the significance of this selected peak. It is measured by the p-value parameter. The statistic which is used in this test has the form:

$$g = \frac{max_{1 \leq i \leq q} I(\omega_i)}{\sum_{i=1}^{q} I(\omega_i)}, \tag{9}$$

where: $I(\omega_i)$ is the spectral estimate (the periodogram) evaluated at Fourier frequencies and $q = [(N-1)/2]$ is the entire of the number $(N-1)/2$. The p-value can be calculated according to the formula [20]:

$$P(g \geq x) = \sum_{j=1}^{p} (-1)^{j-1} \binom{n}{j} (1 - jx)^{n-1} \tag{10}$$

where p is the largest integer less than $1/x$. In our analysis, we use this framework to evaluate a p-value for a given time series of entropy. If we get g^* from the calculation of g, then Equation (10) gives a p-value for probability that the statistic g is larger than g^*.

Proteins can be characterized by various parameters that can be obtained from simulations, such as the earlier mentioned $\langle R \rangle$ or μ. Another meaningful parameter is the RMSD (root mean squared displacement), which can be calculated in every moment of the simulation and is defined as:

$$RMSD = \sqrt{\frac{\sum_{i=1}^{n} R_i \cdot R_i}{n}} \tag{11}$$

where: R_i is a position of the i-th atom. In this way, one can obtain a series of this parameter as a function of time for the further processing. The drawing procedure is realized at the beginning of the simulation from the area of initial values. The initial condition for the albumin molecule drawing is simulated for a given temperature.

We can then group molecules according to mentioned parameters ($\langle R \rangle$, μ, RMSD) and obtain five sets for $300K$, $303K$, $306K$, $309K$, and $312K$, and compare these parameters' values. We decided to use the Kruscal–Wallis statistical test because it does not assume a normal distribution of the data. This nonparametric test indicates the statistical significance of differences in the median between sets. A statistic of this test has the form [21]:

$$H = \frac{12}{N(N+1)} \sum_{i}^{k} \frac{rs_i^2}{n_i} - 3(N+1), \tag{12}$$

where rs_i is a sum of all rank of the given sample, N represents the number of all data in all samples, and k is the number of the groups (set into temperature). We compare the H value of this statistic to the critical value of the χ^2 distribution. If the statistic crosses this critical value, we use a nonparametric multi-comparison test—the Conover–Iman. The formula which we employ has the form [22]:

$$\left| \frac{rs_i}{n_i} - \frac{rs_j}{n_j} \right| > t_{1-\alpha/2} \sqrt{\frac{(S_r - C)(N - 1 - H)}{n_i n_j (N - k)(N - 1)}} \tag{13}$$

where S_r is the sum of all square of ranks, $t_{1-\alpha/2}$ is a quartile from the Student t distribution on $N - k$ degrees of freedom, and C has the value:

$$C = 0.25N(N+1)^2. \tag{14}$$

The multi-comparison test gives the possibility of clustering sets of molecules into those between which there is a statistically significant difference and those between which there is no such difference.

Statistical tests describe differences between medians of the considered sets of molecules. However, they are not informative about differences between distributions of probabilities. In our work, we use the Kullback–Leibler divergence, which is used in information theory, to compare two distributions. The Kullback–Leibler divergence between two distributions p and the reference distribution q is defined as:

$$KL(p,q) = \sum_i p_i \log\left(\frac{p_i}{q_i}\right).$$ (15)

The Kullback–Leibler divergence is also called the relative entropy of p with respect to q. Divergence defined by (15) is not symmetric measure. Therefore, sometimes one can use a symmetric form and obtain the symmetric distance measure [23]:

$$KL_{dist}(p,q) = \frac{KL(p,q) + KL(q,p)}{2}.$$ (16)

The measure defined by (15) is called relative entropy. It is widely used, especially for classification purposes. In our work, we also use this measure to recognize how a given probability distribution is different from a given reference probability distribution. We can look at this method as a complement to the Kruscall–Wallis test. The Kruscall–Wallis test gives information about the differences between medians, while the Kullback–Leibler divergence gives us information about the differences in distributions.

2. Results

We begin the analysis by examining the global dynamics of molecules at different temperatures using the Flory theory, briefly described in the introduction. To estimate the exponent μ from definition in Formula (5), we collect outcomes from a molecular dynamics simulation in the temperature range of $300K$–$312K$. We use a temperature step of $3K$. Such a range of temperature was chosen because we operated within the physiological range of temperatures. During the simulation, we can follow the structural properties of molecules in the considered range of temperatures. In the albumin structure, we can distinguish 22 α-helices, which are presented in Figure 1. Molecules in the given range of parameters display roughly similar structures. Therefore, we chose to study their dynamics in several temperatures. One important property, which we can follow during simulations, is a polymer's end-to-end distance vector as a function of time. We perform the analysis in two time windows, first, between 0–30 ns and, second, between 70–100 ns. For each time window, we obtained the root mean square end-to-end distance of this quantity, whose values are presented in Table 1:

Table 1. Root mean square length of the end-to-end vector for $300K$–$312K$.

Parameter	300K	303K	306K	309K	312K
$\langle R^2 \rangle^{1/2}$ [Å] for 0 ns–30 ns	47.88	46.69	46.72	47.09	47.81
$\langle R^2 \rangle^{1/2}$ [Å] for 70 ns–100 ns	47.12	46.67	47.72	47.13	47.64

We can see that the root mean square length of the end-to-end vector for albumin varies with temperature. To obtain statistics, we can perform the Kruscal–Wallis test of this variation. In our work, values of the statistics in the test are treated as continuous parameters. We do not specify a statistical significance threshold. In Table 2, in the second column, we can see the values of statistics calculated according to Formula (12). The H parameters from the Kruscal–Wallis test have small values for this test, so the results tend to support the hypothesis about equal medians.

Table 2. Values of H statistic in the Kruscal–Wallis test for two time windows for a root mean square end-to-end vector.

Window	Value of H
0–30 ns	$H = 0.62$
70–100 ns	$H = 1.22$

The Kruscal–Wallis test only gives information about differences between medians of the sets, where each set represents one temperature and elements of the set are calculated as root mean square end-to-end vectors over time. In the analysis, for one temperature, we had the set of root mean square end-to-end vectors, and in the next step, we calculated a numerical normalized histogram. Each element of this set corresponds to the single realization of the experiment. The values of the Kullback–Leibler divergence for mean end-to-end signals in the time window 0–30 ns are presented in Table 3. In the first column is the reference distribution for the temperature range. We can see that the biggest value is for the distribution of the temperature $312K$ in the reference distribution $309K$.

Table 3. Kullback–Leibler divergence for root mean square length of mean end-to-end signals in the time window 0–30 ns. The initial data is the same as for the Kruscal–Wallis test.

Temperature of the Reference Probability	300K	303K	306K	309K	312K
300K	0	0.9324	1.3858	0.8946	0.7604
303K	0.9324	0	1.4628	0.8176	1.6165
306K	0.2802	0.3703	0	0.4096	0.1532
309K	0.9062	0.8161	1.4816	0	2.4139
312K	0.2040	0.3710	0.7023	1.1887	0

We can calculate the Kullback–Leibler distance according to Formula (16). The results are presented in Table 4. This measure indicates that, between distributions, the biggest differences are between the distribution for $309K$ and the distribution for $312K$.

Table 4. Symmetric Kullback–Leibler distance for a root mean square length of end-to-end signals in the time window 0–30 ns. The initial data is the same as for the Kruscal–Wallis test.

	300K	303K	306K	309K	312K
300K	0	0.9324	0.8330	0.9004	0.4822
303K	0.9324	0	0.9165	0.8169	0.9938
306K	0.8330	0.9165	0	0.9456	0.4278
309K	0.9004	0.8169	0.9456	0	1.8013
312K	0.4822	0.9938	0.4278	1.8013	0

We can perform the same analyses for the window 70–100 ns. The obtained values of the Kullback–Leibler divergence, which are presented in Table 5. One can see that the biggest difference is between the distribution for $312K$ and the distribution for $303K$. One can see that the maximum has changed and its value is bigger than the maximum for the 0–30 ns window.

Table 5. Kullback–Leibler divergence for a root mean square length of end-to-end signals in the time window 70–100 ns. The initial data is the same as for the Kruscal–Wallis test.

Temperature of the Reference Probability	300K	303K	306K	309K	312K
300K	0	0.1983	0.1983	0.1409	0.2883
303K	0.7474	0	0.8555	1.6675	2.6390
306K	0.7474	0.8555	0	1.6675	1.1576
309K	0.1351	0.4423	0.4423	0	0.1925
312K	0.8113	1.2084	1.8656	0.7532	0

For the symmetric case, results are presented in Table 6. We can see that a symmetrical form of the measure is also between the distribution for $312K$ and the distribution for $303K$.

Table 6. Symmetric Kullback–Leibler distance for a root mean square length of end-to-end signals in the time window 70–100 ns. The data is the same for the Kruscal–Wallis test.

	300K	**303K**	**306K**	**309K**	**312K**
300K	0	0.4728	0.4728	0.1380	0.5498
303K	0.4728	0	0.8555	1.0549	1.9237
306K	0.4728	0.8555	0	1.0549	1.5116
309K	0.1380	1.0549	1.0549	0	0.4728
312K	0.5498	1.9237	1.5116	0.4728	0

Tables 7–10 give the different statistical approaches, without the averaging over the time step). We present there all end-to-end signals for one temperature, and for each moment of time, we calculate the average only over all simulated molecules. In the next step, we calculate normalized histograms. After obtaining numerical distributions, we calculate the Kullback–Leibler divergence between distributions of mean end-to-end signals. The results for the time window 0–30 ns are presented in Table 7.

Table 7. Kullback–Leibler divergence between distributions of mean end-to-end signals in the time window 0–30 ns.

Temperature of the Reference Probability	**300K**	**303K**	**306K**	**309K**	**312K**
300K	0	0.1056	0.3194	0.1367	0.1689
303K	0.0939	0	0.2370	0.1261	0.1425
306K	0.2062	0.1450	0	0.1528	0.1841
309K	0.1276	0.1144	0.1877	0	0.2319
312K	0.1288	0.1275	0.2126	0.1848	0

We can see that the large value of the Kullback–Leibler divergence appears for the distribution for temperature $306K$, where the distribution for temperature $300K$ is a reference distribution. Calculations for the Kullback–Leibler distance for mean end-to-end signals in the time window 0–30 ns are presented in Table 8. The results coincide with the measure of Kullback–Leibler divergence.

Table 8. Kullback–Leibler distance for mean end-to-end signals in the time window 0–30 ns.

	300K	**303K**	**306K**	**309K**	**312K**
300K	0	0.0997	0.2628	0.1321	0.1489
303K	0.0997	0	0.1935	0.1202	0.1350
306K	0.2628	0.1935	0	0.1703	0.1984
309K	0.1321	0.1202	0.1703	0	0.2084
312K	0.1489	0.1350	0.1984	0.2084	0

Following the copula approach presented in [24], we performed the bivariate histograms of some signals to analyse the type of cross-correlation between signals. For the selected example of the mean end-to-end signals in the 0–30 ns time window see Figure 2, while for the 70–100 ns time window, see Figure 3. Observe that for the 0–30 ns case, we have simultaneous "high" events, which refer to the "upper tail dependency". After the simulation time has passed, this dependency is diminished. The copula with the "upper tail dependency", such as the Gumbel one, can provide the proper model of the mean end-to-end signal for the initial simulation time window. For the later simulation time window, we should look for the copula with no "tail dependencies", such as the Frank or Gaussian one.

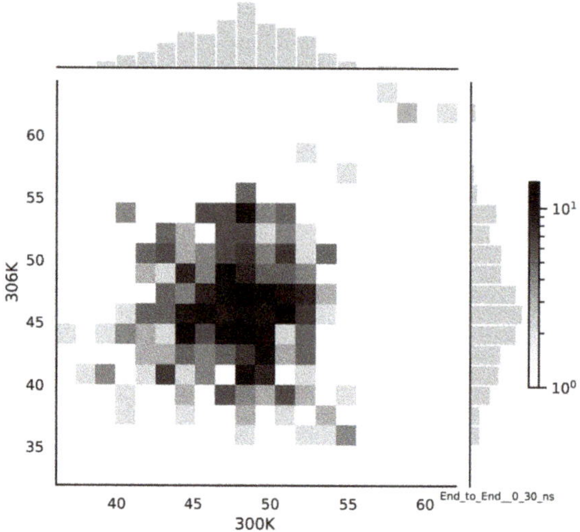

Figure 2. The bivariate histogram of chosen mean end-to-end signals in the 0–30 ns simulation time window.

A similar calculation has been performed for the next time window 70–100 ns. In Table 9, we collect results for Kullback–Leibler divergence for mean end-to-end signals in the time window 70–100 ns.

Table 9. Kullback–Leibler divergence for mean end-to-end signals in the time window 70–100 ns.

Temperature of the Reference Probability	$300K$	$303K$	$306K$	$309K$	$312K$
$300K$	0	0.0920	0.1605	0.1824	0.1183
$303K$	0.1155	0	0.2969	0.1642	0.1474
$306K$	0.1108	0.1324	0	0.1621	0.1206
$309K$	0.1628	0.1266	0.2719	0	0.1206
$312K$	0.0886	0.0979	0.1356	0.1122	0

Its maximal value is between the distribution for $306K$ and the distribution for $303K$ as a reference distribution.

Table 10. Kullback–Leibler distance for mean end-to-end signals in the time window 70–100 ns.

	$300K$	$303K$	$306K$	$309K$	$312K$
$300K$	0	0.1038	0.1356	0.1726	0.1034
$303K$	0.1038	0	0.2147	0.1454	0.1226
$306K$	0.1356	0.2147	0	0.2170	0.1281
$309K$	0.1726	0.1454	0.2170	0	0.1164
$312K$	0.1034	0.1226	0.1281	0.1164	0

In Table 10 we can see that the biggest value of the Kullback–Leibler distance appears between the distributions for $306K$ and $309K$. Comparing this result with the previous one, we see that the maximal Kullback–Leibler distance for the window 0–30 ns changes its place.

In Tables 7–10 the diagonal terms are equal 0, since we are comparing a distribution of end-to-end distributions of raw data for the same time window and temperature. In Table 11 we present selected Kullback–Leibler distances for distributions of raw data end-to-end signals in the time window 70–100 ns and $306K$ temperature.

Table 11. Kullback–Leibler distance for raw data end-to-end signals in the time window 70–100 ns and 306*K*.

Number of Molecule	1	2	3	4	5	6	7	8	9
1	0	0.2927	0.2073	0.2156	0.2450	0.3016	0.1751	0.2072	0.4298
2	0.2927	0	0.1661	0.0919	0.0931	0.1589	0.2367	0.1310	0.2765
3	0.2073	0.1661	0	0.1347	0.2135	0.3059	0.1282	0.1013	0.1843
4	0.2156	0.0919	0.1347	0	0.1203	0.1353	0.1499	0.1338	0.2585
5	0.2450	0.0931	0.2135	0.1203	0	0.1478	0.2635	0.1434	0.3083
6	0.3016	0.1589	0.3059	0.1353	0.1478	0	0.2812	0.1978	0.3122
7	0.1751	0.2366	0.1282	0.1499	0.2635	0.2812	0	0.1318	0.2281
8	0.2072	0.1310	0.1013	0.1338	0.1434	0.1978	0.1318	0	0.1614
9	0.4298	0.2765	0.1843	0.2585	0.3083	0.3122	0.2281	0.1615	0

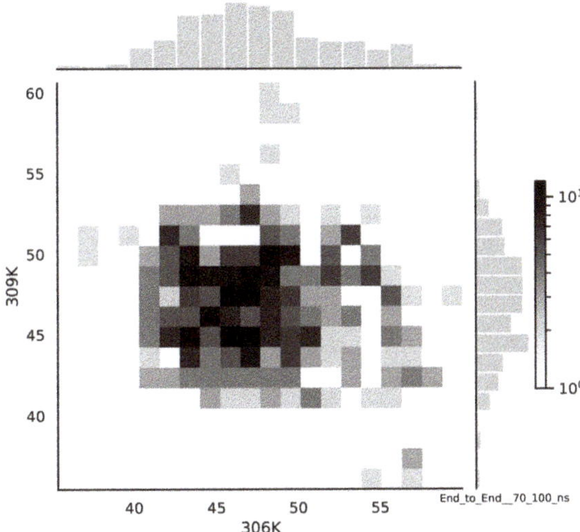

Figure 3. The bivariate histogram of chosen mean end-to-end signals in the 70–100 ns simulation time window.

As two different statistical approaches can not unambiguously distinguish between temperatures (or their subsets), we can conclude that dynamics of the system at this temperatures range is roughly similar. This will be approved by the analysis of the the Flory–De Gennes exponent.

Using Equation (5), we estimate values of the Flory–De Gennes exponent for the given protein chain. Results are presented in Table 12. We can see that the values of the Flory–De Gennes exponents do not differ too much within the examined range of temperature, and they all seem to attain a value of ca. 0.4. The power-law exponents for oligomers span a narrow range of 0.38–0.41, which is close to the value of 0.40 obtained for monomers [25–27].

Table 12. Size exponent obtained according to Formula (5).

Parameter	300*K*	303*K*	306*K*	309*K*	312*K*
Size exponent (μ) for 0 ns–30 ns	0.3986	0.3947	0.3948	0.3960	0.3984
Size exponent (μ) for 70 ns–100 ns	0.3961	0.3946	0.3981	0.3961	0.3978

Detailed values for two time windows are presented in Table 13. We can see that from a global point of view and Flory theory, the situation is quite uniform for all initial conditions and is within the considered range of temperature. Thus, in our modeling, we see an effect of elasticity vs. swelling. This competition is preserved for the examined temperature range.

Table 13. Values of H statistic in the Kruscal–Wallis test for μ parameter for two time windows.

Window	Value of Parameter H
0–30 ns	$H = 0.48$
70–100 ns	$H = 1.47$

The H parameters from the Kruscal–Wallis test have small values for this test, so the results tend to support the hypothesis about equal medians.

The distribution of backbone dihedral angles carries information of the molecule's dynamics, as they are tightly connected to its elasticity. Using simulations, we can determine the angle ϕ and the angle ψ (see Introduction). Next, we can determine energies that depend on these angles by employing Formula (17). In Figures 4 and 5, we present the logarithm of the angle energy component for two temperatures. Figure 4 presents values in a period of time 0–10 ns, and Figure 5 presents values in a period of time 90–100 ns. We can see that these values fluctuate around the mean value. Such dynamics are similar for all temperatures, but the mean energy rises slightly with increasing temperature. In Figure 5, the blue line presents a logarithm of the mean value of angle energy and the gray line presents values of the standard deviation. We can see that changes in energy are much more evident.

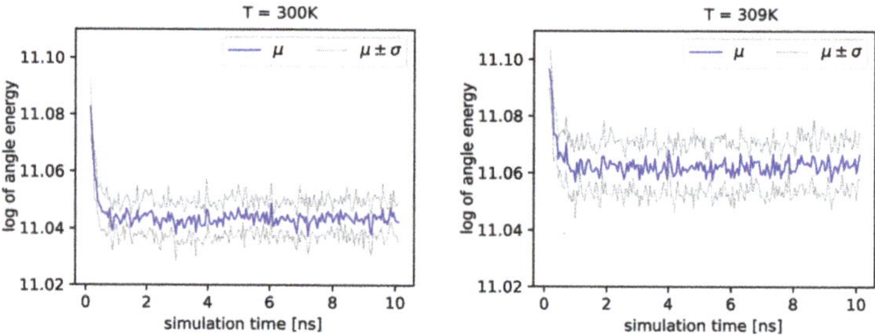

Figure 4. Logarithms of angle energies vs. the simulation time (0–10 ns), exemplary outcome of $T = 300K$ and $T = 309K$.

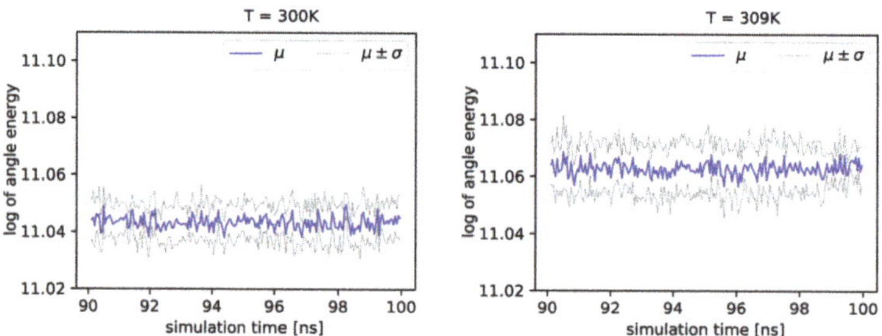

Figure 5. Logarithms of angle energies vs. the simulation time (90–100 ns), exemplary outcome of $T = 300K$ and $T = 309K$.

Different outcomes come from an analysis of the energy component connected with the dihedral angles ψ. See Figure 6 (blue lines) in double logarithmic scale, and note the linear regression lines. We present a period of time from 1 ns to 10 ns.

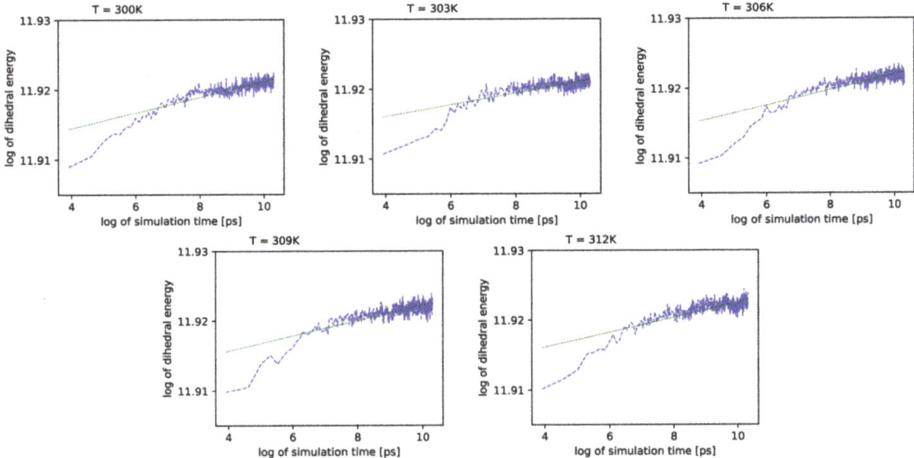

Figure 6. Dihedral energies for each temperature vs. simulation time. We use the double logarithmic scale. Green lines represent linear regression.

Simulations also show that for further times in the log-log scale, these values exhibit almost no change with time.

In Figure 7, we present regression parameters from Figure 6.

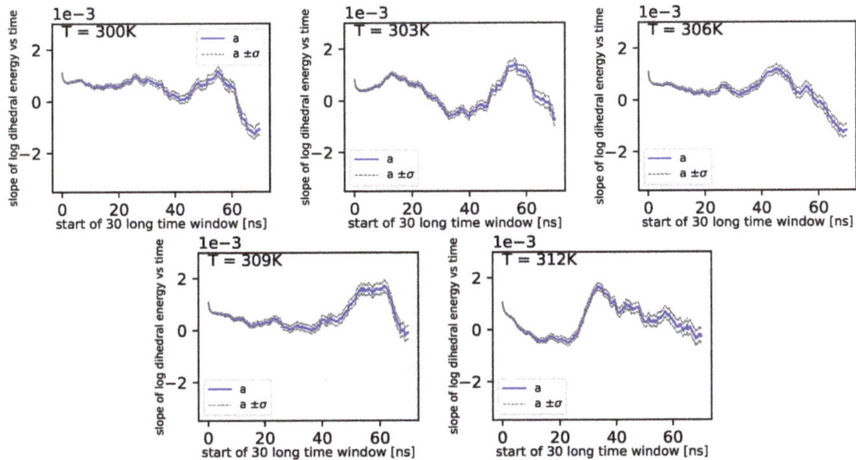

Figure 7. Slope of dihedral angle from regression.

The dynamics of proteins are governed mainly by non-covalent interactions [28]. Therefore, it would be useful to study other components of energy, such as the Coulomb component and the van der Waals component (see Equation (17) and the discussion following it). The Coulomb component oscillates around the almost constant trend present for every temperature. However, the mean value of these oscillations increases with the temperature. The opposite situation appears for the van der Waals component. Similar to the Coulomb component, for each selected temperature, it follows an almost constant trend. However, when the temperature increases, the mean value of such oscillations decreases. The Coulomb component and the van der Waals component obey simple dynamics. Hence, we can move to a further discussion of more interesting dihedral and angular parts of the energy.

Simulations of albumin dynamics can be used to obtain frequency distributions of angles ϕ and ψ. An exemplary 2 D histogram of such a distribution is presented in Figure 8. We can see that most data are concentrated in a small area of space of angles. Similar behavior can be observed for other temperatures and simulation times. This is because the structure of the investigated protein is quite rigid, and there appears to be only a little angular movement. Nevertheless, we can observe variations of the conformational entropy, estimated with such a histogram (see Figure 9).

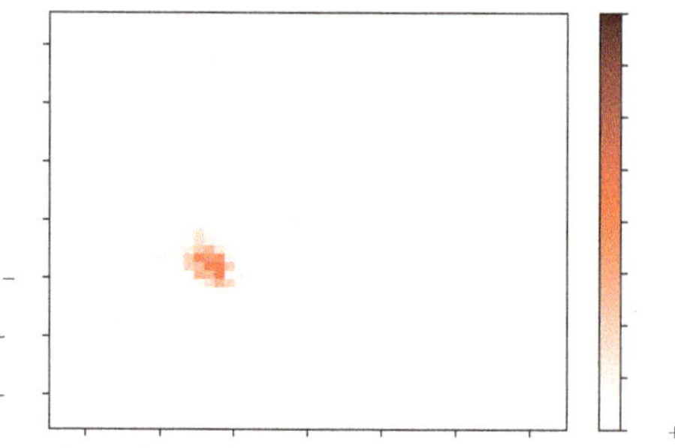

Figure 8. Numerically obtained distribution of angles ϕ and ψ for $T = 300K$ and 30 ns of simulation.

According to formulas connecting the probability distribution of angles and entropy [17], we can follow changes of entropy in time and temperature. Values of conformational entropy for various temperatures and with simulation time 0–30 ns are presented in Figure 9, where the blue line represents the course of the mean entropy value over time. The calculations take into account nine simulations, and one dot presents one arithmetic mean of entropy. Here, we can see that during simulations, some oscillations can appear. Therefore, for each temperature, we perform Fisher's test. Results of this test are presented in Table 14.

Table 14. Probability (p-value) for mean entropy over time in the time interval 0–30 ns.

	300K	303K	306K	309K	312K
p-value	0.849	0.601	0.009	0.536	0.143

A simple comparison of the p-values in Table 14 shows that the lowest value is located around $306K$ degrees. According to Formula (6), mean entropy is supposed to be treated as stochastic, i.e., partially deterministic and partially random. In our work, we treat the p-value as a parameter, which gives us information about the tendency on maintaining the truth of the hypothesis: $\beta = 0$. If the p-value is lower than in another case, then it presents a stronger tendency in favor of the alternative hypothesis: $\beta \neq 0$.

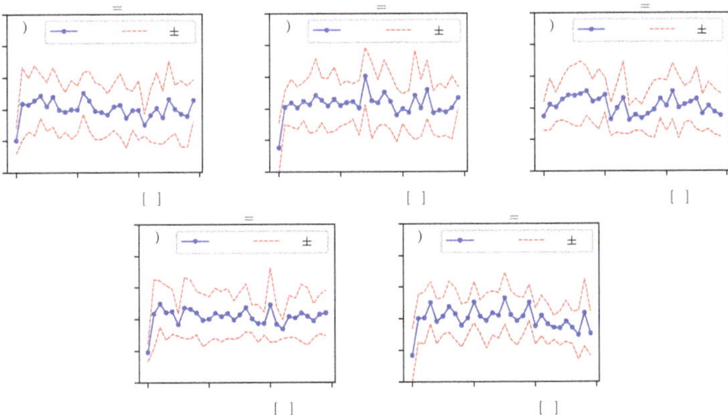

Figure 9. Value of conformational entropy for various temperatures and 0–30 ns of simulation time.

Much more detailed calculations of the p-values for this case are presented in Table 15. We can see that at $306K$, there is a local minimum of the mean p-value. It indicates that at $306K$, the tendency towards the alternative hypothesis: $\beta \neq 0$ is stronger than it is at $303K$ and $309K$. However, for $300K$, the mean p-value parameter is the smallest. The mean p-value parameter is almost the same for $306K$ and $312K$. This means that both temperatures exhibit nearly the same tendency for the hypothesis: $\beta \neq 0$. Such regularities can be seen in Figure 9.

Table 15. Probability (p-value) for entropy of single trajectory over time in the time interval 0–30 ns.

Number of Signal	300K	303K	306K	309K	312K
1	0.439	0.737	0.367	0.754	0.643
2	0.153	0.660	0.961	0.922	0.835
3	0.320	0.483	0.012	0.558	0.0551
4	0.263	0.372	0.700	0.223	0.323
5	0.544	0.657	0.057	0.130	0.066
6	0.114	0.385	0.232	0.416	0.360
7	0.499	0.529	0.008	0.006	0.007
8	0.013	0.871	0.610	0.692	0.456
9	0.245	0.891	0.871	0.832	0.694
mean	0.288	0.621	0.424	0.504	0.382

The same analyzes can be done on the data presented in the image.

The same analyzes can be done on the data presented in Figure 10. The results of these analyzes, in relation to individual time series of entropy, are presented in the Table 16. For these data, we also observe the local minimum of the p-value parameter for $306K$. We have a maximum at $303K$. Thus, the tendency towards the hypothesis $\beta = 0$ is strongest here in comparison to the rest of the analyzed temperatures. This maximum value of the p-value parameter for $303K$ is also present for the previous window.

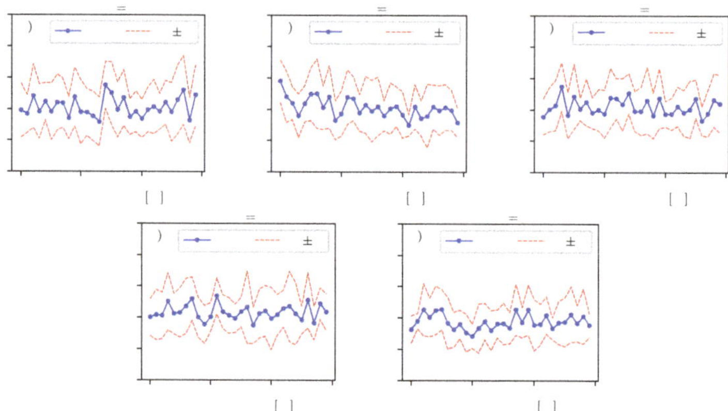

Figure 10. Value of conformational entropy for various temperatures and 70–100 ns of simulation time.

Table 16. Probability (*p*-value) for entropy of a single trajectory over time in the time interval 70–100 ns.

Number of Signal	300K	303K	306K	309K	312K
1	0.328	0.720	0.460	0.078	0.041
2	0.183	0.890	0.146	0.713	0.898
3	0.092	0.378	0.219	0.464	0.264
4	0.764	0.800	0.440	0.629	0.297
5	0.030	0.802	0.275	0.323	0.313
6	0.220	0.108	0.330	0.440	0.540
7	0.279	0.121	0.382	0.253	0.046
8	0.582	0.299	0.976	0.324	0.457
9	0.438	0.550	0.438	0.951	0.642
mean	0.324	0.519	0.407	0.464	0.389

Formula (11) can be calculated for each moment of simulation, so we obtain a time series. To get one parameter for one molecule, we calculate a mean value of this parameter for each molecule. In the next step, we can perform the same statistical considerations to derive values of the H statistic. For the time window 0–30 ns, the value of the Kruscal–Wallis test gives $H = 9.41$; however, for the time window 70–100 ns, the value of this statistic is $H = 12.52$. Because we treat the statistic as an ordinary parameter, we do not specify whether results are significantly different. In our approach, we conclude that the tendency for the alternative hypothesis about not equal medians is much higher for the time window 70–100 ns. Since this tendency is more significant in the second case than in the previous one, we decided to perform a multi-comparison test. We chose the Conover–Iman multi-comparison test. We treated $t_{\alpha-1/2}$ as a parameter, which depends on an assumed statistical significance. Therefore in Table 17, we only give the left side of Equation (13).

Table 17. Results of the Conover–Iman multi-comparison test in the time interval 70–100 ns.

Set of Molecule	300K	303K	306K	309K	312K
300K	–	1.2222	16.1111	12.5556	10.3333
303K	1.2222	–	17.3333	13.7778	11.5556
306K	16.1111	17.3333	–	3.5556	5.7778
309K	12.5556	13.7778	3.5556	–	2.2222
312K	10.3333	11.5556	5.7778	2.2222	–

Here, we have a comparison between medians for different sets. Using values from Table 17 and calculating the right sight of Equation (13) for $p = 0.05$, we can state that there are sets that differ from another cluster of sets. The cluster of sets for temperatures $306K$, $309K$, and $312K$ differ from the set for temperature $303K$. Unfortunately, the set for temperature $300K$ is between these two.

We can compare the above statistical results to the other methods such as the Kullback–Leibler distance, which is presented in the Introduction. In Table 18, we have values of this quantity. We can see that the biggest value is between the distribution for $300K$ and the distribution for $306K$ as a reference distribution.

The calculation of the Kullback–Leibler distance is presented in Table 19. We can conclude that the $306K$ case differs most from all other cases, so the dynamics of the system in the temperature range of $306K$–$309K$ appears to differ most from the dynamics of the system in other temperature ranges.

Table 18. Kullback–Leibler divergence for the mean RMSD in the time window 0–30 ns. The data are the same as for the Kruscal–Wallis test.

Temperature of Reference Distribution	300K	303K	306K	309K	312K
300K	0	0.4045	2.0767	2.917	0.1932
303K	1.6280	0	1.6853	2.4913	0.6891
306K	3.3967	2.6057	0	1.8597	2.3939
309K	2.2053	0.5003	1.1502	0	1.1444
312K	0.1598	0.1343	1.6845	2.5806	0

Table 19. Symmetric Kullback–Leibler distance for the mean RMSD in the time window 0–30 ns. The data are the same as for the Kruscal–Wallis test.

	300K	303K	306K	309K	312K
300K	0	1.0162	2.7367	2.5600	0.1765
303K	1.1016	0	2.1455	1.4958	0.4117
306K	2.7367	2.1455	0	1.5049	2.0392
309K	2.5600	1.4958	1.5049	0	1.8625
312K	0.1765	0.4117	2.0392	1.8625	0

The calculations of Kullback–Leibler divergence and distance for the time window 70–100 ns are presented in Tables 20 and 21. We can conclude that the biggest value is between the distributions for $306K$ and $303K$.

Table 20. Kullback–Leibler divergence for a mean RMSD in the time window 70–100 ns. The data are the same as for the Kruscal–Wallis test.

Temperature of Reference Distribution	300K	303K	306K	309K	312K
300K	0	0.2941	2.7930	2.1527	3.9428
303K	0.8374	0	5.1756	2.2304	3.7127
306K	1.3305	2.4804	0	0.8902	2.8295
309K	4.0173	3.3975	3.3421	0	2.7354
312K	3.6680	2.2372	3.3862	0.8271	0

Table 21. Symmetric Kullback–Leibler distance for the mean RMSD in the time window 70–100 ns. The data are the same as for the Kruscal–Wallis test.

	300K	303K	306K	309K	312K
300K	0	0.5657	2.0617	3.0850	3.8054
303K	0.5657	0	3.8280	2.8139	2.9749
306K	2.0617	3.8280	0	2.1161	3.1078
309K	3.0850	2.8139	2.1161	0	1.7812
312K	3.8054	2.9749	3.1078	1.7812	0

In a further analysis, we compare distributions from signals of the RMSD obtained in the same way as a mean root square of end-to-end signals. Results for the Kullback–Leibler distance are presented in Tables 22 and 23. We can conclude that for the time window 0–30 ns, the most significant value of differences is for the distribution for $309K$ and the distribution for $300K$. For the time window 70–100 ns, the most significant value is for $306K$ and $303K$.

Table 22. Symmetric Kullback–Leibler distance for signals of the RMSD in the time window 0–30 ns.

	300K	**303K**	**306K**	**309K**	**312K**
300K	0	0.3686	0.7090	1.7109	0.1870
303K	0.3686	0	0.3051	1.2909	0.1838
306K	0.7090	0.3051	0	0.7468	0.5706
309K	1.7109	1.2909	0.7468	0	1.4513
312K	0.1870	0.1838	0.5706	1.4513	0

Table 23. Symmetric Kullback–Leibler distance for signals of the RMSD in the time window 70–100 ns.

	300K	**303K**	**306K**	**309K**	**312K**
300K	0	0.1056	6.3487	5.5232	3.9800
303K	0.1056	0	6.4065	5.7737	4.3824
306K	6.3487	6.4065	0	0.9648	1.7829
309K	5.5232	5.7737	0.9648	0	0.4260
312K	3.9800	4.3824	1.7829	0.4260	0

In all tables of the Kullback–Leibler divergence measures we can see 0 value. For all these cases we compare two raw data from the same time window in the same temperature. Results for $306K$ we present in Table 24. We can see that the distribution of Kullback–Leibler distances are widely dispersed. We can observe a similar dispersion in other temperatures.

Table 24. Symmetric Kullback–Leibler distance for raw signals of the RMSD in the time window 70–100 ns for $306K$.

Number of Molecule	1	2	3	4	5	6	7	8	9
1	0	1.669	0.388	0.277	0.137	0.500	3.260	0.560	0.249
2	1.669	0	1.974	2.565	2.074	1.615	0.660	1.498	1.410
3	0.388	1.974	0	0.241	0.617	0.246	3.589	0.299	0.352
4	0.277	2.565	0.241	0	0.250	0.494	3.867	0.628	0.406
5	0.137	2.074	0.617	0.250	0	0.540	3.593	0.597	0.275
6	0.500	1.615	0.246	0.494	0.540	0	3.338	0.080	0.140
7	3.257	0.660	3.589	3.867	3.593	3.337	0	3.118	2.795
8	0.560	1.498	0.300	0.628	0.597	0.080	3.118	0	0.165
9	0.249	1.410	0.352	0.406	0.275	0.140	2.795	0.165	0

Referring to Figure 11, we observe simultaneous small valued events for the 0–30 ns simulation time window. There is no such events for the 70–100 ns simulation time window, see Figure 12. Here, the copula with the "lower tail dependency", such as the Clayton, can provide the proper model of the RMSD for the 0–30 ns simulation time window.

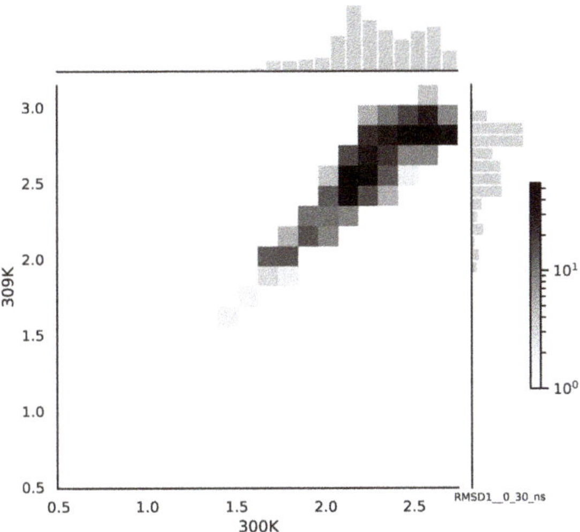

Figure 11. The bivariate histogram of the chosen mean RMSD signals in the 0–30 ns simulation time window.

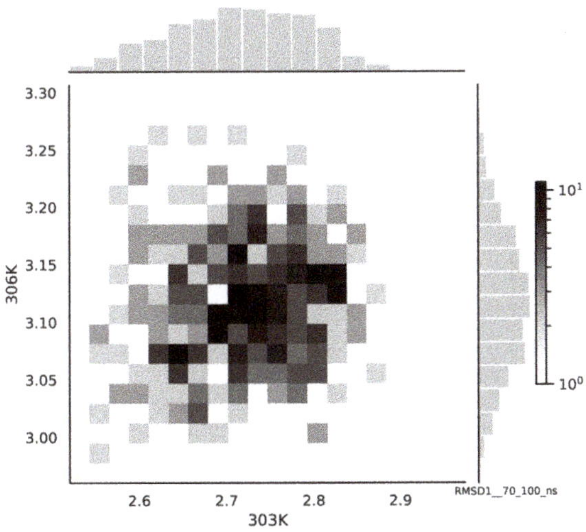

Figure 12. The bivariate histogram of the chosen mean RMSD signals in the 70–100 ns time window.

3. Discussion

Albumin plays an essential role in many biological processes, such as the lubrication of articular cartilage. Hence, knowledge of its dynamics in various physiological conditions can help us to understand better its role in reducing friction. From a general point of view, albumin behaves stable regardless of the temperature. The angle energy of the protein fluctuates similarly regardless of the temperature (see Figures 4 and 5). However, its mean value rises monotonically with temperature. On the other hand, the size exponent μ fluctuates with temperature. We calculate, using the Kruscal–Wallis test, that these fluctuations generate small values of H parameter and, as a result, indicate a hypothesis about the equality of medians.

Further, referring to Figures 6 and 7, we can see that the dihedral energy seems to obey (at least at some range of simulation times) the power law-like relation, although the scaling exponent is relatively small. In Figure 7, we can see this relation in the range of temperature from $306K$–$309K$. This exponent does not change in time t too much. It suggests stable behavior. We consider the conformational entropy of the system. For its plot versus simulation time see Figures 9 and 10. On each graph, we can observe a pattern which seems to attain an oscillatory behavior. Fisher's test allows describing this property by providing the p-value parameter. If this parameter, for some time series of entropy, is relatively small in contrast to p-value parameters for another time series of entropy, then the statistical hypothesis $H_0{:}\beta = 0$ is less probable than in another case. In Table 14, we can observe that in the range of temperature $303K$–$309K$, there is a local minimum of p-values. A similar effect can be observed in Table 15 when we calculate the mean p-value for each column. We can also observe that there is a local minimum of around $306K$. A smaller mean p-value is seen for $300K$ and $312K$. This means that both temperatures exhibit the analogical tendency to favor the hypothesis $\beta \neq 0$. We can also see this in Figure 9, where for temperatures $300K$, $306K$, and $312K$, the regularity of periodicity increases more than at other temperatures. Other tests, including the Kruscal–Wallis test followed by the Conover–Iman multi-comparison test, Kullback–Leibler distance, and Kullback–Leibler divergence give consistent results, while referring to the mean RMSD of the protein of interest.

This globular protein is soluble in water and can bind both cations and anions. By analyzing Figures 1 and 8, it is clear that albumin's secondary structure is mainly standard α-helix. Charged amino acids (AA) have a large share in albumin's structure, especially Aspartic and Glutamic acid (25% of all AA) and hydrophobic Leucine and Isoleucine (16%). This composition enables it to preserve its conformation, due to intramolecular interactions as well as interactions with the solvent. Two main factors are at play here, namely hydrophobic contacts and hydrogen bonds (which also occur between protein and water). A large number of charged AA result in a considerable number of inter- and intramolecular hydrogen bonds. On the other hand, hydrophobic AA result in a more substantial impact when the hydrophobic effect stabilizes a protein's core. As shown by Rezaei-Tavirani et al. [29], the increase of temperature in the physiological range of temperatures results in conformational changes in albumin, which cause more positively charged molecules to be exposed. This, in turn, results in a lower concentration of cations near a molecule. Albumin is known to largely contribute to the osmotic pressure of plasma, where the presence of ions largely influences this property. An increase of temperature results in a reduction of the osmotic pressure of blood, which in feverish conditions can lead to a higher concentration of urine. Because water is the main component of blood and it has a high heat capacity, the increase of urine concentration results in better removal of heat from the body [29]. The difference in dynamics between temperatures shown in the present study could indicate albumin's binding affinity with other SF components and thus changes its role in biolubrication toward anti-inflamatory. However, due to the high complexity of the fluid, more research has to be performed.

These chemical properties result in an effect of elasticity vs. swelling competition for albumin chains immersed in water. If the elasticity of the chain fully dominates over albumin's swelling-induced counterpart (interactions of polymer beads and water with network/bond creation), the exponent would have a value of $1/2$. If, in turn, a reverse effect applied, the value of the exponent should approach the value of $1/4$, [30] (note the two first Eqs only). As a consequence, and what has not been discovered by any other study, the exponent obtained from the simulation in the examined range of temperature looks as if it is near an arithmetic mean of the two exponents mentioned, namely: $\mu = 1/2(2/4 + 1/4) = 3/8$, thus pointing to $0.375 \sim 0.4$. The difference of about 0.025 very likely comes from the fact that the elastic effect on albumin shows up to win over the swelling-assisted counterterm(s), cf. the Hamiltonian used for the simulations. The exponent $\mu = 3/8$ is for itself called the De Gennes exponent, and it is reminiscent of the De Gennes dimension-dependent gelation exponent $2/(d + 2)$, see [30] (eq 15 therein; $d = 3$-case). It suggests that albumin's elastic effect, centering at about 0.4, is fairly well supported by the internal network and thus supports the creation of bonds, which has been revealed by the present study. Of course, the overall framework is well-substantiated in terms of the scaling concept [3,7]. Regular oscillations of entropy suggest that the system oscillates around the equilibrium state. Furthermore, the Flory–De Gennes exponent appears to be unchanged for both observation windows. Therefore, we expect the system to be near the equilibrium state. We expect the dynamic of the system, for longer simulation times, to be similar to the presented one. The AMBER force field has been reported on overstabilizing helices in proteins. Therefore our next work will report on the effect of force field and water model used.

4. Materials and Methods

The structure of human albumin serum (code 1e78) has been downloaded from a protein data bank (https://www.rcsb.org/structure/1E78) as a starting point to simulations. We use the YASARA Structure Software (Vienna, Austria) [31] to perform MD simulations. Besides this, a three-site model (TIP3P) of water was used [32]. All-atom simulations were performed under the following conditions: temperature $310K$, $pH = 7.0$ and in 0.9% NaCl aqueous solution, with a time step of $2fs$. Berendsen barostat and thermostat with a relaxation time of $1fs$ were used to maintain constant temperature and pressure. To minimize the inuence of rescaling, YASARA does not use the strongly fluctuating instantaneous microscopic temperature to rescale velocities at each simulation step (classic Berendsen thermostat). Instead, the scaling factor is calculated according to Berendsen's formula from the time average temperature. The charge was −15. Simulations were carried for 100 ns, due to the fact that throughout this time very small fluctuations of entropy measured could be seen. All-atom molecular dynamics simulations were performed using the AMBER03 force field [33]. The AMBER03 potential function describing interactions among particles takes into account electrostatic, van der Waals, bond, bond angle, and dihedral terms:

$$E_{total} = \sum_{bonds} k_b \left(r - r_{eq} \right)^2 + \sum_{angle} k_\phi \left(\phi - \phi_{eq} \right)^2$$
$$+ \sum_{dihedrals} \frac{V_n}{2} [1 + \cos(n\psi - \gamma)] + \sum_{i<j} \left[\frac{A_{ij}}{r_{ij}^{12}} - \frac{B_{ij}}{r_{ij}^{6}} + \frac{q_i q_j}{\epsilon r_{ij}} \right] \quad (17)$$

where, k_b and k_ϕ are the force constants for the bond and bond angles, respectively; r and ϕ are bond length and bond angle; r_{eq} and ϕ_{eq} are the equilibrium bond length and bond angle; ψ is the dihedral angle and V_n is the corresponding force constant; the phase angle γ takes values of either $0°$ or $180°$. The non-bonded part of the potential is represented by van der Waals (A_{ij}) and London dispersion terms (B_{ij}) and interactions between partial atomic charges (q_i and q_j). ϵ is the dielectric constant. For each spherical angle distribution, we performed a normalized histogram consisting of 50 bins. Such a bin number is a compromise between a resolution and smoothness of histograms. Taking the empirical probability p_i of data being in i-th bin, we estimate the entropy [18,34]:

$$S = -R_0 \sum_i p_i \log(p_i), \quad (18)$$

where we use the gas constant $R_0 = 8.314 \frac{J}{K \cdot mol}$. The sum goes over the discretized Ramachandran space. Obviously, since entropy is an information measure, it depends on a bin size, hence we have here only an estimation applicable for entropy comparisons for different temperatures and simulation times, since bin sizes are constant for all simulations and temperatures. Numerical data analyzes were performed in the Scilab development environment [35].

5. Conclusions

We present several scenarios of an analysis of simulations of the albumin dynamics. The albumin is the linear protein, which makes simulations and analysis straightforward. In particular, we get the end-to-end vector values, and we present that changes for exponent do not vary significantly. The more detailed analysis can be performed, however, using the entropy that appears to oscillate in the function of the simulation time. In general, these oscillations are regular, which is approved by employing the statistical test.

However, our findings show that near the $306K$ temperature, we can observe a local minimum of the p-value of this test. Interestingly, this minimum changes its position near the $312K$ temperature for the 70–100 ns time window. We also consider the RMSD parameter, which for the second time window, exhibits significant differences between the median of the sets defined for temperatures—we observe clustering.

When summarizing, let us note that the elastic energy terms included in the Hamiltonian (Equation (17)) provide a particularly robust bond vs. angle contribution to the elastic energy of the biopolymer, especially when arguing its role in the angles' domain. The remaining non-elastic terms in Equation (17), in turn, contribute, in general, to the swelling assisted co-effect. Thus, the already mentioned Van der Waals and electrostatic terms are responsible for the overall attraction–repulsion behavior of the network-like swollen albumin, whereas the last "liquid–crystalline" type of energetic inclusion to Equation (13) samples the dihedral angles space, an observation already discussed above.

Author Contributions: Conceptualization, P.W., P.B., K.D., and A.G.; investigation, P.W., P.B., K.D., and D.L.; writing—original draft preparation, P.W. and P.B.; writing—review and editing, P.B., K.D., D.L. and A.G. All authors have read and agreed to the published version of the manuscript.

Funding: This work is supported by UTP University of Science and Technology, Poland, grant BN-10/19 (P.B. and A.G.).

Acknowledgments: We wish to thank Maryellen Zbrozek, E.T.A. at UTP (Fulbright fellowship) for her kind help in assessing the final form of our paper.

Conflicts of Interest: The authors declare no conflict of interest.

References

1. Grimaldo, M.; Roosen-Runge, F.; Zhang, F.; Schreiber, F.; Seydel, T. Dynamics of proteins in solution. *Quarter. Rev. Biophys.* **2019**, *52*, e7. [CrossRef]
2. Rubinstein, M.; Colby, R.H. *Polymer Physics*; Oxford University Press: Oxford, UK, 2003.
3. De Gennes, P.G. *Scaling Concepts in Polymer Physics*; Cornell University Press: Ithaca, NY, USA, 1979.
4. Gadomski, A. On (sub) mesoscopic scale peculiarities of diffusion driven growth in an active matter confined space, and related (bio) material realizations. *Biosystems* **2019**, *176*, 56–58. [CrossRef] [PubMed]
5. Metzler, R.; Jeon, J.H.; Cherstvy, A.G.; Barkai, E. Anomalous diffusion models and their properties: Non-stationarity, non-ergodicity, and ageing at the centenary of single particle tracking. *Phys. Chem. Chem. Phys.* **2014**, *16*, 24128. [CrossRef] [PubMed]
6. Weber, P.; Bełdowski, P.; Bier, M.; Gadomski, A. Entropy Production Associated with Aggregation into Granules in a Subdiffusive Environment. *Entropy* **2018**, *20*, 651. [CrossRef]
7. Flory, P. *Principles of Polymer Chemistry*; Cornell University Press: Ithaca, NY, USA, 1953.
8. Doi, M.; Edwards, S.F. The Theory of The Polymer Dynamics. In *The International Series of Monographs on Physics*; Oxford University Press: New York, NY, USA, 1986; pp. 24–32.
9. Bhattacharya, A.A.; Curry, S.; Franks, N.P. Binding of the General Anesthetics Propofol and Halothane to Human Serum Albumin. *J. Biol. Chem.* **2000**, *275*, 38731–38738. [CrossRef] [PubMed]
10. Oates, K.M.N.; Krause, W.E.; Jones, R.L.; Colby, R.H. Rheopexy of synovial fluid and protein aggregation. *J. R. Soc. Interface* **2006**, *3*, 167–174. [CrossRef] [PubMed]
11. Rebenda, D.; Čípek, P.; Nečas, D.; Vrbka, M.; Hartl, M. Effect of hyaluronic acid on friction of articular cartilage. *Eng. Mech.* **2018**, *24*, 709–712.
12. Seror, J.; Zhu, L.; Goldberg, R.; Day, A.J.; Klein, J. Supramolecular synergy in the boundary lubrication of synovial joints. *Nat. Commun.* **2015**, *6*, 6497. [CrossRef]
13. Katta, J.; Jin, Z.; Ingham, E.; Fisher, J. Biotribology of articular cartilage—A review of the recent advances. *Med. Eng. Phys.* **2000**, *30*, 1349–1363. [CrossRef]

14. Moghadam, M.N.; Abdel-Sayed, P.; Camine, V.M.; Pioletti, D.P. Impact of synovial fluid flow on temperature regulation in knee cartilage. *J. Biomech.* **2015**, *48*, 370–374. [CrossRef]

15. Liwo, A.; Ołdziej, S.; Kaźmierkiewicz, R.; Groth, M.; Czaplewski, C. Design of a knowledge-based force field for off-latice simulations of protein structure. *Acta Biochim. Pol.* **1997** *44*, 527–548. [CrossRef]

16. Havlin, S.; Ben-Avraham, D. Diffusion in disordered media. *Adv. Phys.* **2002**, *51*, 187–292. [CrossRef]

17. Bhattacharjee, S.M.; Giacometti, A.; Maritan, A. Flory theory for polymers. *J. Phys. Condens. Matter* **2013**, *25*, 503101. [CrossRef] [PubMed]

18. Baxa, M.C.; Haddadian, E.J.; Jumper, J.M.; Freed, K.F.; Sosnick, T.R. Loss of conformational entropy in protein folding calculated using realistic ensembles and its implications for NMR-based calculations. *Proc. Natl. Acad. Sci. USA* **2014** *111*, 15396–15401. [CrossRef]

19. Ahdesmaki, M.; Lahdesmaki, H.; Yli-Harja, O. Robust Fisher's test for periodicity detection in noisy biological time series. In Proceedings of the IEEE International Workshop on Genomic Signal Processing and Statistics, Tuusula, Finland, 10–12 June 2007.

20. Wichert, S.; Fokianos, K.; Strimmer, K. Identifying periodically expressed transcripts in microarray time series data. *Bioinformatics* **2004**, *20*, 5–20. [CrossRef] [PubMed]

21. Kanji, G.K. *100 Statiatical Tests*; SAGE Publications Ltd.: Thousand Oaks, CA, USA, 2006.

22. Conover, W.J. *Practical Nonparametric Statistics*, 2nd ed.; John Wiley & Sons: New York, NY, USA, 1980.

23. Gupta, A.; Parameswaran, S.; Lee, C.-H. Classification of electroencephalography signals for different mental activities using Kullback-Leibler divergence. In Proceedings of the IEEE International Conference on Acoustics, Speech and Signal Processing, Taipei, Taiwan, 19–24 April 2009.

24. Domino, K.; Błachowicz, T.; Ciupak, M. The use of copula functions for predictive analysis of correlations between extreme storm tides. *Physica A* **2014**, *413*, 489–497. [CrossRef]

25. Tanner, J.J. Empirical power laws for the radii of gyration of protein oligomers. *Acta Crystallogr. D Struct. Biol.* **2016**, *72*, 1119–1129. [CrossRef]

26. Korasick, D.A.; Tanner, J.J. Determination of protein oligomeric structure from small-angle X-ray scattering. *Protein Sci.* **2018**, *27*, 814–824. [CrossRef]

27. De Bruyn, P.; Hadži, S.; Vandervelde, A.; Konijnenberg, A.; Prolič-Kalinšek, M.; Sterckx, Y.G.J.; Sobott, F.; Lah, J.; Van Melderen L.; Loris, R. Thermodynamic Stability of the Transcription Regulator PaaR2 from *Escherichia coli* O157:H7. *Biophys. J.* **2019**, *116*, 1420–1431. [CrossRef]

28. Durell, S.R.; Ben-Naim, A. Hydrophobic-Hydrophilic Forces in Protein Folding. *Biopolymers* **2017**, *107*, e23020. [CrossRef]

29. Rezaei-Tavirani, M.; Moghaddamnia, S.H.; Ranjbar, B.; Amani, M.; Marashi, S.A. Conformational Study of Human Serum Albumin in Pre-denaturation Temperatures by Differential Scanning Calorimetry, Circular Dichroism and UV Spectroscopy. *J. Biochem. Mol. Biol.* **2006** *39*, 530–536. [CrossRef]

30. Isaacson, J.; Lubensky, T.C. Flory exponents for generalized polymer problems. *J. Phys. Lett.* **1980**, *41*, L-469–L-471. [CrossRef]

31. Krieger, E.; Vriend, G. New ways to boost molecular dynamics simulations. *J. Comput. Chem.* **2015**, *36*, 996–1007. [CrossRef] [PubMed]

32. Mark, P.; Nilsson, L. Structure and Dynamics of the TIP3P, SPC, and SPC/E Water Models at 298 K. *J. Phys. Chem. A* **2001**, *105*, 9954–9960. [CrossRef]

33. Duan, Y.; Wu, C.; Chowdhury, S.; Lee, M.C.; Xiong, G.; Zhang, W.; Yang, R.; Cieplak, P.; Luo, R.; Lee, T.; et al. A point-charge force field for molecular mechanics simulations of proteins based on condensed-phase quantum mechanical calculations. *J. Comput. Chem.* **2003**, *24*, 1999–2012. [CrossRef] [PubMed]

34. Baruah, A.; Rani, P.; Biswas, P. Conformational Entropy of Intrinsically Disordered Proteins from Amino Acid Triads. *Sci. Rep.* **2015**, *5*, 11740. [CrossRef]

35. SciLab. Available online: https://www.scilab.org (accessed on 30 April 2017).

Article

Applications of Information Theory Methods for Evolutionary Optimization of Chemical Computers

Jerzy Gorecki

Institute of Physical Chemistry, Polish Academy of Sciences, Kasprzaka 44/52, 01-224 Warsaw, Poland; jgorecki@ichf.edu.pl

Received: 15 February 2020; Accepted: 8 March 2020 ; Published: 10 March 2020

Abstract: It is commonly believed that information processing in living organisms is based on chemical reactions. However, the human achievements in constructing chemical information processing devices demonstrate that it is difficult to design such devices using the bottom-up strategy. Here I discuss the alternative top-down design of a network of chemical oscillators that performs a selected computing task. As an example, I consider a simple network of interacting chemical oscillators that operates as a comparator of two real numbers. The information on which of the two numbers is larger is coded in the number of excitations observed on oscillators forming the network. The parameters of the network are optimized to perform this function with the maximum accuracy. I discuss how information theory methods can be applied to obtain the optimum computing structure.

Keywords: chemical computing; oscillatory reaction; genetic optimization; classification problem; interacting oscillators

1. Introduction

The domination of semiconductor technology in modern information processing comes from the fact that man-made semiconductor logic gates are reliable, fast and inexpensive. Their mean time of trouble-free work is probably longer than a human life time. The technology allows combining the logic gates together, leading to complex information processing devices. Therefore, the dominant information coding is based on the binary representation and the design uses the bottom-up strategy, allowing to make more complex information processing devices as a combination of the simple ones [1]. Such an approach is highly successful, leading to exaFLOPS calculations, cloud computing or the fast internet.

Information processing in living organisms is based on chemical reactions [2]. Although, in the case of numerical calculations, the chemical information processing is many orders of magnitude slower than that achieved in silicon circuits, the dedicated chemicals computers (read brains) can execute many algorithms faster than a powerful modern electronic computer. This remark applies to problems that require three-dimensional reconstruction of space and mapping complex motion. Most humans can learn to drive a car, whereas the algorithms for autonomous cars are still far from being applicable. Chemical information processing in living organisms is usually reliable during an animal life. However, the activity time of chemical information processing media used in experiments, as for example, the Belousov–Zhabotinsky (BZ) reaction, is measured in hours rather than in years. Therefore, the bottom-up approach to the construction of chemical information processing devices is not useful for potential experimental verification of theoretical concepts because of the short lifetime of the components.

There are many strategies in which a chemical medium can be applied for information processing. Information can be coded in concentrations of reagents, in the spatial structures or in the spatio-temporal evolution. An excitable chemical medium allows for easy realization of logic gates [3–7]. In such gates, the input and output states are coded in the presence or absence of an excitation at a selected point of the computing medium within a specific time interval. However, the most effective algorithms are obtained if the chemical information processing medium works in parallel. For example, it happens if reactions at different points are coupled by the diffusion. Two classical algorithms of reaction–diffusion computing belong to such class. One of them is the prairie-fire algorithm, which allows finding the shortest path in the labyrinth using wave propagation in an excitable medium [8,9]. The other is the Kuhnert algorithm for image processing with the light inhibited variant of the oscillatory BZ-reaction [10–12].

The BZ-reaction has probably been the most studied chemical reaction where the nonlinear phenomena are clearly manifested [13,14]. The reaction is an oxidation of an organic substrate by bromine compounds in an acidic environment and in the presence of a catalyst. The BZ-reaction became famous because oscillations can be easily observed, as the changes in concentrations of the catalyst in different oxidation forms are reflected by the medium color. If the ferroine is used as the catalyst then the medium is red when the reduced catalyst ($Fe(phen)_3^{2+}$) is dominant and the medium becomes blue for a high concentration of the catalyst in the oxidized form ($Fe(phen)_3^{3+}$). The reaction includes an autocatalytic production of the reaction activator ($HBrO_2$). If the medium is spatially distributed and if the diffusion of the activator is allowed, then the region corresponding to a high concentration of the activator can trigger the reaction around and a pulse of the activator propagating in space can appear. The interest in a BZ-reaction as a medium for chemical information processing comes from the fact that its properties are similar to those observed for the nerve system [2]. Using a spatially distributed medium, one can form channels where propagation of excitation pulses is observed. These pulses interact (annihilate) one with another and can change their frequency on non-excitable junctions between channels [15]. The output information is usually coded in the presence of an excitation pulse at a given point and at a specific time. The successes of reaction–diffusion computing with an excitable medium seem to confirm the key role of excitability in biological information processing [16]. Perhaps this is true, but the recent results suggest that an oscillatory medium and information coded in the number of oscillations can also be efficiently applied for chemical computing.

If the ruthenium complex ($Ru(bpy)_3$) is used as the catalyst, then the BZ reaction is photosensitive and illumination with the blue light produces Br^- ions that inhibit the reaction [17–19]. After illumination of such an oscillatory medium, excitations are rapidly damped and the system reaches a stable, steady state. On the other hand, the oscillatory behavior re-appears immediately after the illumination is switched off [20]. The existence of external control is very important for information processing applications because it allows inputting information into the computing medium [10–12]. For the analysis presented below, it is sufficient to assume that the controlling factor has an inhibiting effect. In the following part of the paper, following the analogy with the photosensitive BZ-reaction, I use the word illumination to describe the control factor.

Our recent results suggest that reasonably accurate database classifiers can be constructed with a network of coupled chemical oscillators [21–23]. The database records are assumed to have a form of predictors followed by the record type. The classification algorithm is supposed to return a correct record type if the predictor values are known. In this approach, it is assumed that each oscillator in the network can be individually inhibited by an external factor. We can use this factor to introduce the input information and to control the evolution of the medium. Oscillators in the considered information processing networks belong to two types. There are input oscillators and their illuminations are related to predictor values of a given record. There are also so called "normal" oscillators and their illuminations are fixed. The normal oscillators are supposed to moderate interactions in the medium and optimize them for a specific problem. Therefore, the locations of normal oscillators and their illuminations define the program executed by the network. It is also assumed that the output information about the record type can be extracted from the number of excitations (the number of maxima of a specific reactant) observed at a selected set of oscillators within a fixed interval of time. In such an approach, information processing is a transient phenomenon. It does not matter if the system approaches a stationary state over a long time or not.

It has been demonstrated [21–23] that the top-down approach can be successfully applied to design classifiers based on coupled chemical oscillators. Within the top-down approach, we first specify the function that should be performed by the considered system. Next, we search for possible factors that can modify the system evolution and increase its information processing ability. Finally, we combine all these factors and apply them to achieve the optimum performance. The top-down design reflects the idea of Evolution, where the struggle to survive is the goal, but the measures to achieve this goal and possible synergy between different factors increasing the fitness of an individual organism are not fully understood. Having in mind that the structure of the brain is a product of evolution, it is not surprising that the evolutionary optimization can be a useful tool for finding the values of parameters describing a computing medium that performs a specific operation. The evolutionary optimization [24,25] works on a population of classifiers. The best classifiers are allowed to recombine their parameters and to produce an offspring that is included in the next generation. Spontaneous mutations of classifier parameters are also taken into account. As a result of trial and error, the fitness of the best classifier increases with the number of generations [21–23]. Of course, it can happen that the considered medium is completely useless for the task we like to perform. Nevertheless, even in cases where the selected medium is needless for a solution of the considered information processing task, the application of evolutionary optimization should allow us to estimate the medium usefulness.

In the following, I investigate the classifier corresponding to the problem of which of the two real numbers $x, y \in [0, 1]$ is larger as an example application of information theory to chemical computing. The methods of information theory suggest the type of fitness function for the evolutionary optimization of a classifier formed by coupled chemical oscillators. The motivation for the research is to illustrate that chemical computing can be efficient if we optimize the computing medium and give some margin for potential errors. Actually, this is something that characterizes information processing in the living organisms ("errare humanum est"). For the verification, if $y > x$, a network of oscillators that solves this problem with reasonable accuracy can be quite simple, and as shown below, it can be made of just three oscillators.

2. Results

The problem of which of the two numbers $x, y \in [0, 1]$ is larger has a direct geometrical interpretation. A pair (x, y) represents a point in the unit square. The problem if $y > x$ is equivalent to determining if the corresponding point is located above the unit square diagonal (cf. Figure 1a).

Chemistry suggests many strategies that allow us to verify which of two numbers is larger. For example, on can consider a reaction:

$$X + Y \rightarrow products$$

in which one molecule of X is consumed together with a single molecule of Y. One can start such a reaction with initial concentrations of [X] and [Y] equal to x and y, respectively. After a long time, the only remaining molecules are from the reagent that was the majority: there are the molecules of X if $x > y$ and molecules of Y if the reverse relation was held. So spectroscopy should give us an answer to which of the two numbers x, y is larger.

The same strategy applies if we use a more complex reaction, as for example:

$$aX + bY + Z \rightarrow products$$

but in such a case, we obtain the relationship between the values of ax and by.

The problem of the relationship between two numbers x, y can also be solved with an oscillatory reaction that is inhibited by light. It is known that the period of the photosensitive BZ-reaction increases with illumination [26]. The numbers x, y can be translated into light intensities $I(x)$ and $I(y)$ using the function $I(q) = \alpha * q + \beta$ ($\alpha, \beta > 0$) such that for $q \in [0, 1]$ the values of $I(q)$ are in the range corresponding to rapid changes in the period. Next, we need to apply illuminations $I(x)$ and $I(y)$ to two identical reactors and measure the periods. The oscillator characterized by a longer period was illuminated by light with intensity corresponding to the larger of two numbers x, y.

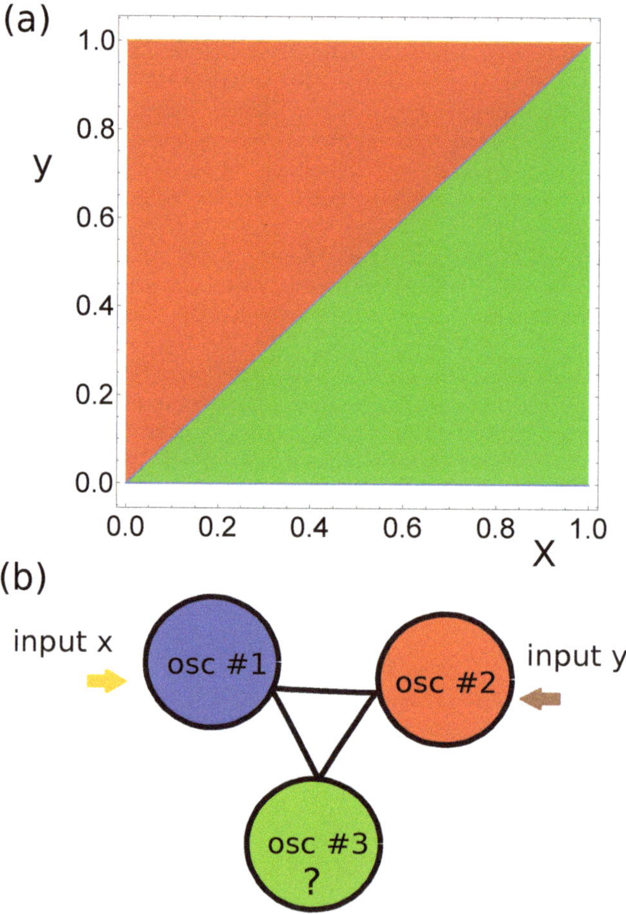

Figure 1. (**a**) The geometrical interpretation of the problem of which of the two numbers is larger. The areas $y > x$ and $y < x$ are colored red and green respectively, (**b**) I postulate that the problem of which of the two numbers is larger can be solved with the illustrated network of three coupled oscillators. Having in mind the symmetry of the problem, I assume that oscillators #1 and #2 are inputs of x and y. The role of oscillator #3, the network parameters and location of the output oscillators are determined by the evolutionary optimization.

Here, I consider yet another strategy of solving the problem of which of the two numbers is larger by formulating it as a database classification. Problems of database classification are quite general because many algorithms can be re-formulated as database classifications. Let us consider an algorithm

A that returns a discrete output on an input in the form of N real numbers belonging to the algorithm domain (D_A). Moreover, we assume that the number of possible answers is finite. Let $(K+1) \in \mathbb{N}$ describe the number of answers. Formally, we describe such an algorithm as a map:

$$A : \mathbb{R}^N \supset D_A \to O, \tag{1}$$

where $O = \{0, 1, ..., K\}$. If we consider an element $(p_1, ..., p_N) \in D_A$, then:

$$A : D_A \ni p = (p_1, ..., p_N) \mapsto t_p \in \{0, 1, ..., K\}. \tag{2}$$

Let us introduce a set $E_A(L)$, which contains L arguments of the algorithm A:

$$E_A(L) = \{p^s = (p_1^s, p_2^s, ..., p_N^s); p^s \in D_A \wedge s \in \{1, 2, ..., L\}\}. \tag{3}$$

We can generate the database $F_A(L)$ made of records constructed in the following way:

$$F_A(L) = \{(p_1^s, p_2^s, ..., p_N^s, t_{p^s}); (p_1^s, p_2^s, ..., p_N^s) \in E_A(L) \wedge t_{p^s} = A(p^s)\}. \tag{4}$$

Each record of $F_A(L)$ contains N predictors $p_1^s, p_2^s, ..., p_N^s$ and the record type t_{p^s}. The classifier of the database $F_A(L)$ is supposed to return the correct record type if the predictor values are used as the input. A classifier that correctly classifies any dataset $F_A(L)$ can be seen as a program that executes the algorithm A. For example, the XOR operation

$$XOR : \{0, 1\} \times \{0, 1\} \to \{0, 1\}$$

can be completely described by the classification of database:

$$F_{XOR}(4) = \{(0, 0, 0), (0, 1, 1), (1, 0, 1), (1, 1, 0)\}.$$

Therefore, the chemistry based classifier of $F_{XOR}(4)$ is a chemical realization of the XOR gate. The same approach applies to any logical operations involving those of multivariable ones.

Within the formalism presented above, the problem which of two numbers is larger can be seen as an algorithm $A_>$:

$$A_> : [0, 1] \times [0, 1] \ni (x, y) \mapsto t_{(x,y)} \in \{0, 1\},$$

where $t_{(x,y)} = 1$ if $x \geq y$ and $t_{(x,y)} = 0$ iff $x < y$. Therefore, the problem can be formulated as the classification problem for databases in the form:

$$F_>(L) = \{(x^s, y^s, t_{(x^s, y^s)}); (x^s, y^s) \in [0, 1] \times [0, 1] \wedge s \in \{1, 2, ..., L\}\}$$

I postulate that the problem of $F_>(L)$ database classification can be solved by a network of three coupled oscillators in the geometry illustrated in Figure 1b. I assume that the output information is coded in numbers of excitations observed on oscillators forming the network. Such a method of extracting the output is motivated by the fact that a chemical counter of excitation number can be easily constructed [27], so the information read-out can be done with chemistry.

2.1. The Time Evolution Model of an Oscillator Network

In this section, I briefly introduce oscillator networks, discuss the specific properties of the Belousov–Zhabotinsky reaction that can be useful for construction of classifiers and introduce a simple model of network time evolution. The detailed information can be found in [28].

The networks of chemical oscillators can be formed in different ways. One can use individual continuously stirred chemical reactors (CSTRs) and link them by pumps ensuring the flow of reagents [29,30]. Alternatively, networks of oscillators can be formed by touching droplets containing a water solution of reagents of an oscillatory BZ reaction stabilized by lipids dissolved in the surrounding oil phase. If phospholipids (asolectin) are used, then BZ droplets communicate mainly via an exchange of the reaction activator ($HBrO_2$ molecules) that can diffuse through the lipid bilayers and transmits excitation between droplets [31]. A high uniformity of droplets that form the network can be achieved if droplets are generated in a microfluidic device [32]. One can also use DOVEX beads or silica balls [33] to immobilize the catalyst and inhibit oscillations by illumination or electric potential [34].

The simplest mathematical model of a BZ-reaction describes the process as an interplay between two reagents: the activator and the inhibitor (the oxidized form of the catalyst). Two variable models, such as the Oregonator [35] or Rovinsky–Zhabotinsky model [36], give a pretty realistic description of simple oscillations, excitability and the simplest spatio-temporal phenomena. However, the numerical complexity of models based on kinetic equations is still substantial, and they are too slow for large scale evolutionary optimization of a classifier made as a network of oscillators. Here, following [28], I use the event based model. I assume that three different phases: excited, refractive and responsive can be identified during a single oscillation cycle of a typical chemical oscillator [21,22,28]. The excited phase denotes the peak of activator concentration. A chemical oscillator in the excited phase is able to spread out the activator molecules and trigger excitations in the medium around. In the refractory phase, the concentration of the inhibitor is high, and an oscillator in this phase does not respond to an activator transported from oscillators around. In the responsive phase, the inhibitor concentration decreases. An oscillator in the response phase can get excited by interactions with an oscillator in the excited phase. Following the previous papers [28], the oscillation cycle combines the excitation phase lasting 1 s, the refractive phase, lasting 10 s and the responsive phase that is 19 s long (cf. Figure 2). For an isolated oscillator, the excitation phase appears again after the responsive phase ends and the cycle repeats. Thus, the period of the cycle is 30 s. Oscillations with such a period have been observed in experiments with BZ-medium [20]. The separation of oscillation cycle into phases allows introducing a simple model for interactions between individual oscillators. An oscillator in the refractory phase does not respond to the excitations of its nearest neighbors. An oscillator in the responsive phase can be activated by an excited neighbor. I also assume that if an oscillator is in the excitation phase, then 1 s later, all oscillators coupled with it, that are in the responsive phase, switch into the excitation phase. It is also assumed that after illumination is switched on, the phase changes into the refractory one. When the illumination is switched off, the excited phase starts immediately. The model defined above is much faster than models based on the kinetic equations.

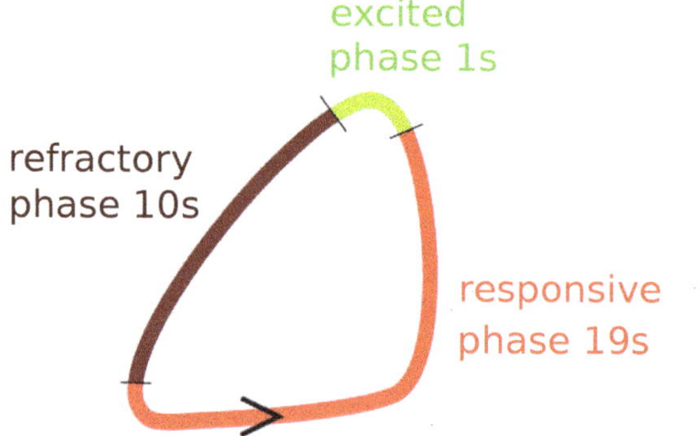

Figure 2. Graphical illustration of the event-based model used to describe the time evolution of an oscillator. Three phases: excited (green), refractory (brown) and responsive (red) follow one after another. The arrow marks the direction of time. Numbers give lengths of corresponding phases.

2.2. The Computing Medium Made of Interacting Oscillators and Its Evolutionary Optimization

In this section, I introduce the parameters needed for numerical simulations of a chemical classifier. In order to use a network of coupled oscillators for information processing, we have to specify the time interval $[0, t_{max}]$ within which the time evolution of the network is observed. This is an important assumption because it reflects the fact that the information processing is a transient phenomenon. I assume that output information can be extracted by observing the system within the time interval $[0, t_{max}]$, and it is not important if the network reaches a steady state before t_{max} or not.

It is assumed that the state of each oscillator in the network can be controlled by an external factor (illumination). There are two types of oscillators in the network: the normal ones and the input ones. For a normal oscillator k, its activity is inhibited within the time interval $[0, t_{illum}^{(k)}]$ ($0 \le t_{illum}^{(k)} \le t_{max}$). After the time $t_{illum}^{(k)}$ the oscillation cycle starts from the excited phase. For a given classification problem the set of times $t_{illum}^{(k)}$ defining the inhibition of normal oscillators is the same for all processed records. This set of times defines the "program" executed by the network to solve the problem. If an oscillator is considered as an input for the j-th predictor and if the predictor value is $p_j \in [0, 1]$, then such oscillator is inhibited (illuminated) within the time interval $[0, t_{start} + (t_{end} - t_{start}) * p_j]$, where the values of t_{start}, t_{end} (both $< t_{max}$) are the same for all predictors. The network geometry (i.e., locations of input and normal oscillators and the geometry of interactions between oscillators) together with illuminations of normal oscillators $t_{illum}^{(k)}$ and times t_{start} , t_{end} fully define the network and allow for simulations of its time evolution after the predictor values are selected. In the following, I will assume that the classifier output is represented by the number of excitations observed on oscillators within the time interval $[0, t_{max}]$.

To verify if the network performs its classification function correctly, I introduce the testing database $F_>(L = 100,000)$ composed of records of the form $(x, y, t_{(x,y)})$ where x, y are uniformly distributed random numbers from $[0, 1]$. The classifier accuracy is calculated as the fraction of correct answers for records from $F_>(L = 100,000)$. As the zero-th order approximation, one can neglect observation of the classifier evolution and say that x is always larger than y. If $F_>(L = 100,000)$ is selected without bias, then such classifier should show 50% accuracy.

In order to increase the accuracy, we have to optimize the classifier parameters. In the optimization discussed below, they included the maximum time within which the system evolution is observed (t_{max}), the locations of input oscillators and the oscillators characterized by fixed illuminations (the normal ones), the times defining the introduction of input values (t_{start} and t_{end}) and the illuminations of normal oscillators $t_{illum}^{(k)}$. Moreover, defining the classifier, we should decide how the output is extracted from the evolution. I select two strategies. I postulate that the classifier output can be read out from the number of excitations on a selected oscillator or as a pair of numbers of excitations observed on two selected oscillators. The quality of a classifier C_A corresponding to the algorithm A can be estimated if one decides how to read the output information but without interpreting if the obtained result corresponds to the case $x \ge y$ or $x < y$. To do it, we can use the mutual information between the set of record types and the set of classifier outputs. Let us consider a set of arguments $E_A(L)$ (Equation (3)) and the related database $F_A(L)$ (Equation (4)) of the algorithm A. Let us assume that the classifier C_A produces the output string a_{p^s} on the record $p^s \in E_A(L)$:

$$C_A : E_A(L) \ni p^s \mapsto a_{p^s}$$

Now let us consider three lists:

$$B = \{a_{p^s}; p^s \in E_A(L)\}$$

$$O = \{t_{p^s}; (p^s, t_{p^s}) \in F_A(L)\}$$

and

$$OB = \{(t_{p^s}, a_{p^s}); (p^s, t_{p^s}) \in F_A(L)\}$$

The mutual information between B and O defined as [37]:

$$MI(O, B) = H(O) + H(B) - H(OB) \tag{5}$$

where $H(X)$ is the Shannon information entropy of the strings that belong to the list X. If this list contains k different strings then:

$$H(X) = -\sum_k r(k) * \log(r(k)) \tag{6}$$

where $r(k)$ is the probability of finding the string #k in the list.

In order to calculate $MI(O, B)$, one needs to specify how the output is read out of the network evolution. For example, if the number of excitations observed on a single oscillator within the time interval $[0, t_{max}]$ is used as the output string, then we should consider all oscillators as potential candidates for the output one. The oscillator, for which the maximum $MI(O, B)$ is achieved, is selected as the output one. In the case of the output coded in a pair of excitation numbers observed on two selected oscillators, one should consider all pairs of oscillators as candidates for the output oscillators. Like in the case of a single oscillator, the pair that produces that maximum $MI(O, B)$ is considered as the output. The maximum value of $MI(O, B)$ obtained within a given method of extracting the output string is considered as the quality (fitness) of the classifier C_A.

Let us notice that when $a_{p^s} = t_{p^s}$, so the classifier works without errors, then $H(O) = H(B) = H(OB)$ and $MI(O, B) = H(O)$. On the other hand, if the answers a_{p^s} are not correlated with the record types t_{p^s} then $H(OB) = H(O) + H(B)$ and $MI(O, B) = 0$. In general, $MI(O, B)$ is an average number of bits of the information about the string t_{p^s} we get if we know the string a_{p^s}. Therefore, we can expect that an increase in $MI(O, B)$ does reflect the rise in classifier accuracy.

To define a classifier, we need to specify: *Geom*—the network geometry and interactions between oscillators, *Loc*—location of the input oscillators, t_{max}, t_{end}, t_{start} and $t_{illum}^{(k)}$. In the considered problem of verification of which of the two numbers is larger, I assumed that the classifier had a triangular geometry, as illustrated in Figure 1b, and that all oscillators were interconnected. In such a geometry, all oscillators are equivalent. Therefore, we can assume that oscillator #1 is the input of the first predictor. Moreover, the system symmetry allows us to assume that oscillator #2 is the input of the second predictor. Therefore, the only missing element of the classifier structure is the type of oscillator #3 (Osc_3) and it is a subject of optimization. I did not introduce any constraints on the type of this oscillator. It could be an input oscillator or the normal one.

The optimization of all parameters of the classifier (Osc_3, t_{max}, t_{end}, t_{start}, $t_{illum}^{(3)}$) was done using the evolutionary algorithm. The technique has been described in detail in [28]. At the beginning, 1000 classifiers with randomly initialized parameters were generated. Of course, it would be naive to believe that a randomly selected network of oscillators performs an accurate classification of the selected database. The generated networks made the initial population for evolutionary optimization. Next, the fitness (Equation (5)) of each classifier was evaluated. In order to speed up the algorithm, the fitness was calculated using the database $F_>(L = 10,000)$ where predictors were $10,000$ randomly generated pairs (x, y). The database $F_>(L = 10,000)$ was 10 times smaller than $F_>(L = 100,000)$ used to estimate the accuracy of the optimized classifier. However, $F_>(L = 10,000)$ is still large enough to contain a representative number of records characterizing the problem. The upper 10% of the most fitting classifiers were copied to the next generation. The remaining 90% of classifiers of the next generation were generated by recombination and mutation processes applied to pairs of classifiers randomly selected from the upper 50% of the most fitting ones. At the beginning, recombination of two parent classifiers produces a single offspring by combining randomly selected parts of their parameters. Next, the mutation operations were applied to the offspring. It was assumed that oscillator

#3 can change its type with the probability $p_{type} = 0.1$. Moreover, random changes of t_{max}, t_{end}, t_{start} and $t_{illum}^{(k)}$ were also allowed. The evolutionary procedure was repeated more than 500 times. A typical progress of optimization is illustrated in Figures 3a and 5a.

2.3. Chemical Algorithms for Verification of Which of the Two Numbers Is Larger

In this section, I discuss the optimized networks that verify which of the two real numbers x and y from [0,1] is larger.

At the beginning, let us consider the network that produces the output as a number of excitations observed at a selected node. The progress of optimization on a small database with 1000 records is illustrated in Figure 3a. The optimization continued for 500 evolutionary steps. For the optimized network $t_{max} = 130.55$ s. As it has been assumed, oscillators #1 and #2 correspond to inputs of x and y, respectively. The values of t_{start} and t_{end} are 69.54 and 23.79 s, respectively. The highest mutual information was obtained if oscillator #3 was the normal one and $t_{illum}^{(3)} = 49.55$ s. Figure 3b illustrates the numbers of cases corresponding to different numbers of excitations and the relationship between x and y for records from $F_>(L = 100,000)$. The red and green bars correspond to $y > x$ and $y < x$, respectively. Figure 3b compares the number of cases observed on different oscillators of the network. It is clear that oscillator #3 is useless as the output because it has excited the same number of time by inputs with $y > x$ and by inputs with $y < x$. On the other hand, for oscillators #1 and #2, the situation is very different. For example, three excitations are observed mainly for $y > x$, whereas four or five excitations are dominant for $y > x$. Therefore, we can define the classifier output as the number of excitations on oscillator #1 such that three excitations correspond to $y > x$ and four or five excitations to $y \leq x$. The accuracy of such a classifier is 82.4%. Figure 4 illustrates the location of correctly and incorrectly classified pairs $(x, y) \in [0, 1] \times [0, 1]$ if the output reading rule defined above is used. The correctly classified pairs in which $y > x$ and $y < x$ are marked by red and green points, respectively. Incorrectly classified pairs in which $y > x$ and $y < x$ are marked by blue and black points, respectively. Similar results are obtained when oscillator #2 is used as the output (Figure 3b), but the accuracy is 82.2%. I believe the small difference between using oscillators #1 and #2 as the output is related to randomness in the generated database.

In order to increase the classification accuracy, I considered the same network but another rule of reading out the information. I assumed that the output is coded in a pair of excitation numbers observed on two oscillators. The initial progress of optimization with $F_>(L = 10,000)$ is illustrated in Figure 5a. The optimization procedure was continued for 1000 generations. The classifier structure was the same as in the previous case; oscillator #3 was the normal one. The optimized classifier is characterized by $t_{max} = 500.00$ s, $t_{start} = 498.95$ s, $t_{end} = 0.15$ s and $t_{illum}^{(3)} = 71.42$ s. The numbers of excitations observed on oscillators #1 and #2 were used as the classifier answer. The translation of the observed number of excitations into the classifier answer is illustrated in Figure 5b. The mutual information between the classifier outputs and the record types in $F_>(L = 100,000)$ is 0.962 and the classification accuracy is 98.2%. Both numbers are remarkably high, especially having in mind that the computing medium is formed by three oscillators only. The optimization procedure has shown that the classifier accuracy strongly depends on t_{max}, as illustrated in Figure 5c. In the optimization procedure, the value of t_{max} was limited at 500s. Figure 6 illustrates the location of correctly and incorrectly classified pairs $(x, y) \in [0, 1] \times [0, 1]$ if the output reading rule defined in Figure 5b is used.

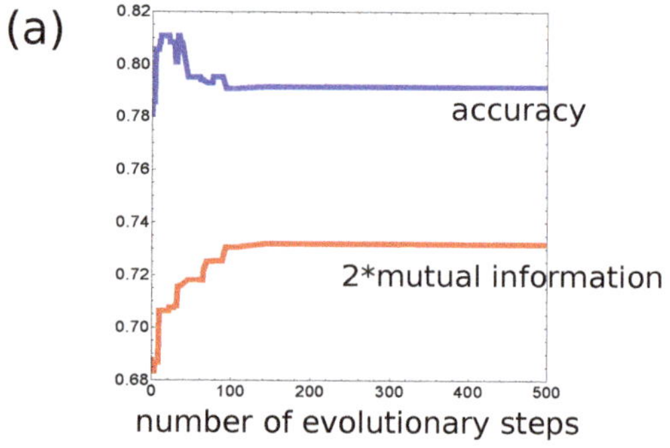

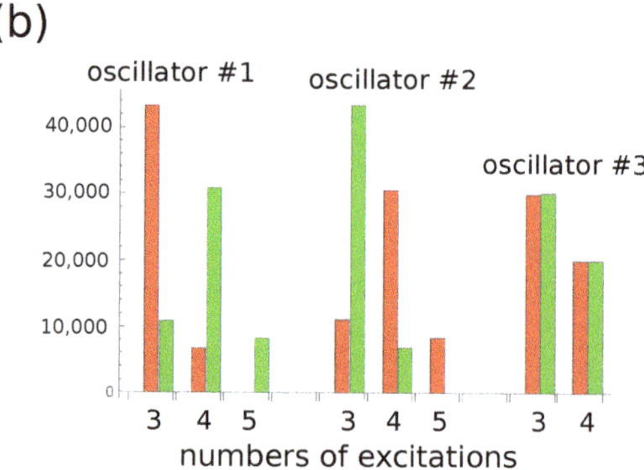

Figure 3. (**a**) The mutual information as a function of an evolutionary step for the classifier in Figure 1b if the number of excitations observed on a single oscillator is used as the output, (**b**) the numbers of cases corresponding to different numbers of excitations and for different relationships between x and y. The records from $F_>(L = 100,000)$ were used. The red and green bars correspond to $y > x$ and $y < x$, respectively. The number of cases recorded on different oscillators are shown.

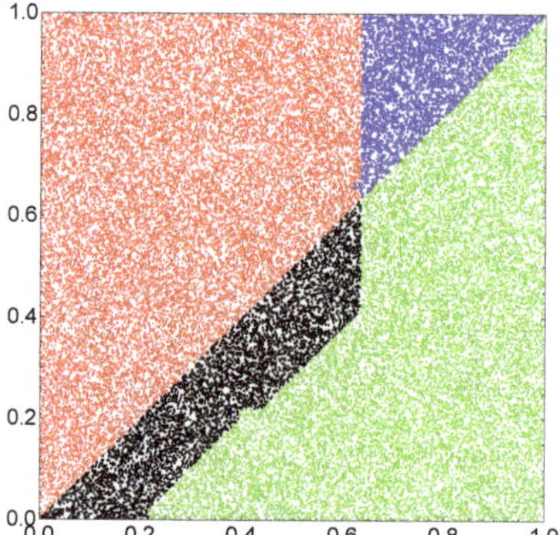

Figure 4. The location of correctly and incorrectly classified pairs $(x, y) \in [0, 1] \times [0, 1]$ if the number of excitations observed on a single oscillator is used as the output. Correctly classified pairs in which $y > x$ and $y < x$ are marked by red and green points, respectively. Incorrectly classified pairs in which $y > x$ and $y < x$ are marked by blue and black points, respectively.

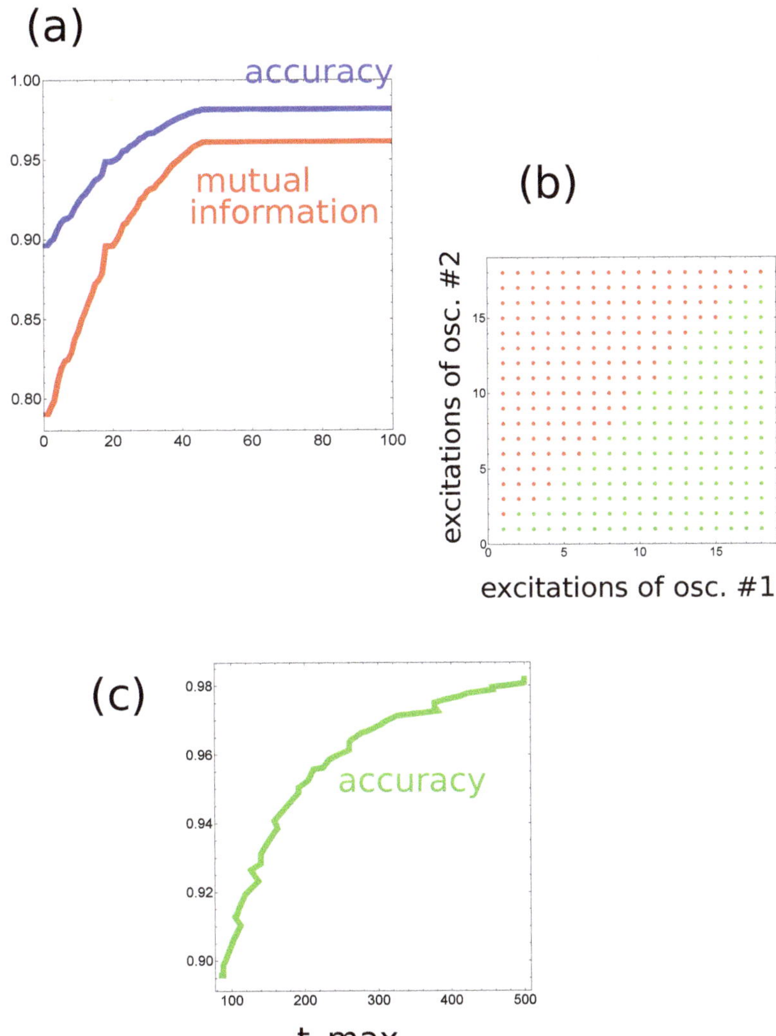

Figure 5. (a) The mutual information as the function of an evolutionary step for the classifier in Figure 1b if the pair of excitation numbers observed on two selected oscillators is used as the output. (b) The table that translates the numbers of excitations observed on oscillators #1 and #2 into the classifier answer. The red and green points correspond to $y > x$ and $y < x$, respectively. (c) The increase in classifier accuracy as a function of t_{max} observed during the optimization.

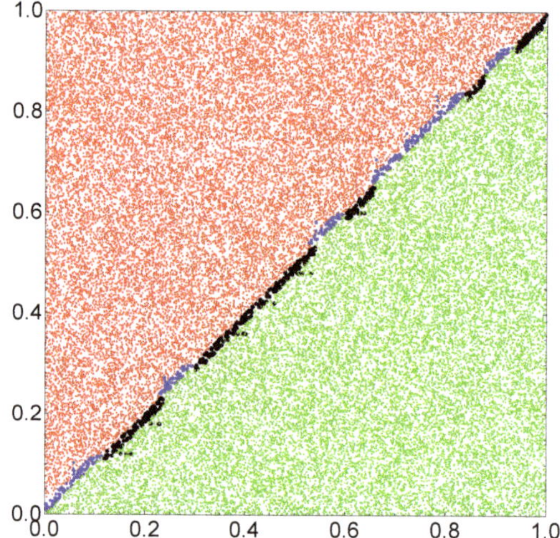

Figure 6. The location of correctly and incorrectly classified pairs $(x,y) \in [0,1] \times [0,1]$ if the pair of excitation numbers observed on oscillators #1 and #2 is used as the output. Correctly classified pairs in which $y > x$ and $y < x$ are marked by red and green points, respectively. Incorrectly classified pairs in which $y > x$ and $y < x$ are marked by blue and black points, respectively.

2.4. Shadows on Optimization Towards the Maximum Mutual Information

As in the previous papers [21–23,28], I assumed that the increase in mutual information means that the classifier accuracy increases, so one can use easy-to-calculate mutual information to estimate the quality of a classifier. But does the increase in the mutual information really mean that the classifier accuracy is higher? The example given below illustrates that it is not.

Let us consider the following example. Assume a database with $2N$ records corresponding to two types v and w of the form:

$$F = \{(p^s, v); 1 \leq s \leq N\} \cup \{(p^s, w); N+1 \leq s \leq 2N\}. \tag{7}$$

Therefore, the list of record types $O = \{t_{p^s}; (p^s, t_{p^s}) \in F\}$ contains N elements v and N elements w and the Shannon information entropy $H(O) = 1$.

Let us also assume that there is a classifier C of the database F that produces two outputs a_1 and a_2 in the following way: If the record type is v, then for L ($L < N/2$) records, the classifier output is a_2, and for $N - L$ records, the classifier output is a_1. Let us redistribute these database records such that: $a_{p^s} = a_2$ if $1 \leq s \leq L$ and $a_{p^s} = a_1$ if $L+1 < s \leq N$. Moreover, if the record type is w, then for M ($M < N/2$) records, the classifier output is a_1, and for $N - M$ records, the classifier output is a_2. Yet again let us redistribute these database records such that: $a_{p^s} = a_1$ if $N+1 \leq s \leq N+M$ and $a_{p^s} = a_2$ if $N + M < s \leq 2N$. Therefore, it is more likely to get the a_1 answer if the record type is v and the a_2 answer if the record type is w. The interpretation: the record type is v if the classifier produces a_1, and the record type is w if the classifier produces a_2 leads to the highest accuracy equal to:

$$Accuracy(k,l) = 1 - (k+l)/2, \tag{8}$$

where $l = L/N$ and $k = K/N$. The contour plot of $Accuracy(k,l)$ is shown in Figure 7a.

The list of classifier answers B on the database F is made of $N - L + M$ symbols a_1 and $N - M + L$ symbols a_2. Therefore;

$$H(B)(k,l) = -(0.5 - k/2 + l/2)\log_2(0.5 - k/2 + l/2) - (0.5 + k/2 - l/2)\log_2(0.5 + k/2 - l/2) \tag{9}$$

For the considered classifier, the list OB is formed of L strings (v, a_2), $N - L$ strings (v, a_1), K strings (w, a_1) and $N - K$ strings (v, a_2). The Shannon information entropy $H(OB)$ equals:

$$H(OB)(k, l) = -k/2 \log_2(k/2) - (0.5 - k/2) \log_2(0.5 - k/2) - l/2 \log_2(l/2) - (0.5 - l/2) \log_2(0.5 - l/2) \qquad (10)$$

and the mutual information between record types and answers of the classifier is:

$$MI(k, l) = 1 + H(B)(k, l) - H(OB)(k, l) \qquad (11)$$

The function $MI(k, l)$ is illustrated in Figure 7b.

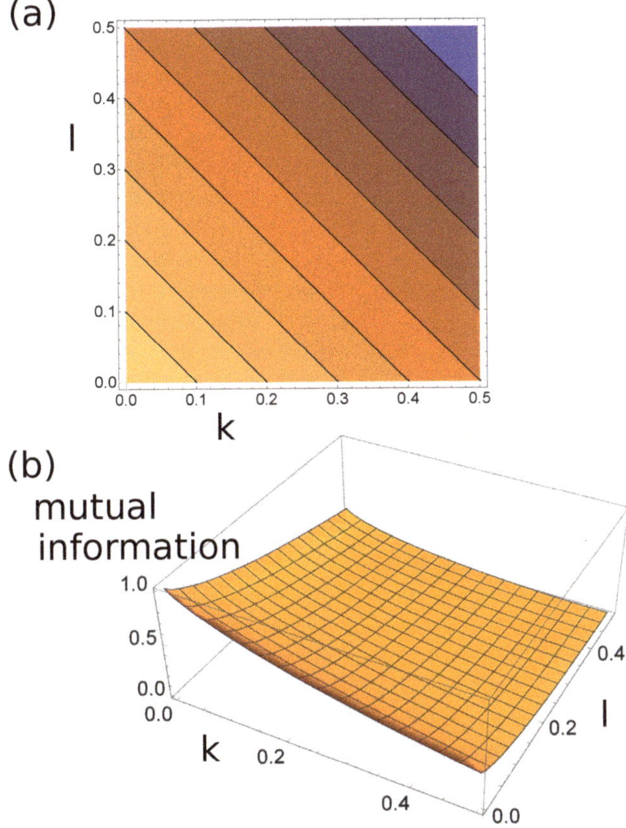

Figure 7. (a) The countour plot of $Accuracy(k, l)$ and (b) the mutual information $MI(k, l)$ for the classifier discussed in Section 2.4 as functions of k and l. The contour lines in (a) are equally distributed between 0.5 and 1.0.

Comparing the dependence of $Accuracy(k, l)$ with $MI(k, l)$, we can identify a path in (k, l) space along which one of the functions increases and the other decreases. An example of such a path is marked by the red line in Figure 8a. The values of $Accuracy(k, l)$ and $2 * MI(k, l)$ along this path are shown in Figure 8b. As seen, the increase in $MI(k, l)$ can slightly decrease $Accuracy(k, l)$.

Having this result in mind, we can say that although optimization based on the mutual information is attractive because it can be easily applied in an optimization program and does not involve direct translation of the evolution into the classification result, a better classifier can still be evolved if its accuracy is used to estimate the classifier fitness.

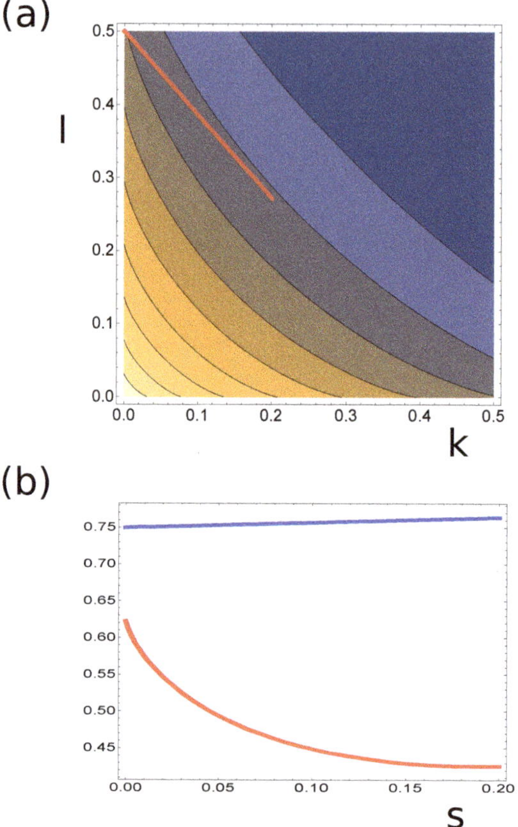

Figure 8. (**a**) The countour plot of the mutual information illustrated in Figure 7b with a hypothetical optimization path shown in red. (**b**) The accuracy (blue curve) and the double of mutual information (red curve) along the path marked in (**a**).The value $s = 0$ describes the point $(0, 0.5)$.

3. Discussion: Why can a Small Network of Chemical Oscillators So Accurately Determine Which of the Two Numbers Is Larger?

The computing networks described above were obtained by a computer program without any human help except of formulating the problem. Now, having the problem solved, we may think over why such high accuracy of classification of which of the two numbers is larger has been achieved.

Let us notice that the problem of which of the two numbers (x, y) is larger can be approximated by solving the problem if $x \geq 0.5$. At a first glance, the problems seem quite different. However, if the pairs are randomly distributed in the cube $[0, 1] \times [0, 1]$ (see Figure 1a), then if $x_1 \geq 0.5$, then with 75% probability, x_1 is the larger number in a pair (x_1, y). And similarly, if $x_2 < 0.5$, then with 75% probability, x_2 is the smaller number in the pair (x_2, y). Construction of a device that checks if a number is smaller or larger than 0.5 using a single chemical oscillator is very easy. Let us assume that the chemical oscillator has a period T. Let us consider a classifier made of a single oscillator. This oscillator serves both as the input of x and the output. Let us consider the classifier is characterized by $t_{max} = 2T - \epsilon$, $t_{start} = t_{max}$ and $t_{end} = 0$. For the input value z, the input illumination is $t_{input}(z) = t_{start} + (t_{end} - t_{start}) * z = (2 * T - \epsilon) * (1 - z)$. If $z < 1 - T/(2 * T - \epsilon)$, then $t_{input}(z) > T - \epsilon$ so only one excitation of the oscillator is observed in the time interval $[t_{input}(z), t_{max}]$. On the other hand, if $z \geq 1 - T/(2 * T - \epsilon)$, then $t_{input}(z) \leq T - \epsilon$ and two excitations are observed. Therefore, such a single oscillator classifier correctly predicts if the input is larger than 0.5 for most of numbers from the interval $[0, 1]$. If we use it to compare two numbers, then using the arguments presented above, we obtain an accuracy of 75%. The idea of a single oscillator classifier presented above can be easily generalized to verify if the input z belongs to one of subintervals of $[0, 1]$. Let us assume that $t_{max} = nT - \epsilon$, $t_{start} = t_{max}$ and $t_{end} = 0$. For the input value z, the input illumination is $t_{input}(z) = t_{start} + (t_{end} - t_{start}) * z = (n * T - \epsilon) * (1 - z)$. Therefore, the oscillator can exhibit $1 + floor(z * n)$ different numbers of oscillations (assuming the we can neglect small ϵ). Precisely j oscillations are observed for $t_{input}(z) \in (\max(0, (n - j)T - \epsilon), (n - j + 1)T - \epsilon]$, which corresponds to $z \in [1 - ((n - j + 1) * T - \epsilon)/(nT - \epsilon), 1 - ((n - j) * T - \epsilon)/(nT - \epsilon))$. Therefore, by observing the number of excitations on a single oscillator, we can estimate the input value with the accuracy $\pm 1/(2n)$.

Now let us come back to the problem of which of the two numbers in a pair (x, y) is larger. Consider a classifier made of two non-interacting oscillators. One of these oscillators works as an input of x and another as an input of y, as described above. Next, we compare the number of excitations on both oscillators. If the number of excitations on the oscillator acting as x input is larger, then $x > y$. If the number of excitations is smaller, then $y > x$ (cf. Figure 5b). If the numbers of excitations are equal, then we can assume, for example, $x > y$ because for uniformly distributed data it will give 50% of correct results. Therefore, the accuracy of such a classifier can be estimated as $1 - 1/(2n)$. Such a trend can be seen in Figure 5c; thus, the evolutionary algorithm was able to discover the idea of the classifier presented above. For $t_{max} = 500$ and $T = 30$, the value of n is 17, so we would expect the classifier error around 0.029. The error of the optimized classifier, based on three interacting oscillators, was smaller and equal to 0.018.

4. Conclusions and Perspectives

In the paper, I discussed how information theory methods can be applied for optimization of a classifier using a network of interacting oscillators as a chemical medium. The problem has been inspired by the fact that bottom-up design of chemical computers is inefficient. It usually produces large structures requiring a precise fit of reaction parameters [38]. The structures resulting from bottom-up design can be difficult for experimental realization. In this aspect, the methods inspired by information theory provide tools for top-down design of compact and reasonably accurate chemical processing media. The use of mutual information for estimating the quality a of chemical information processing medium seems effective because it does not require specification of the rule used to translate the output information into the classification result. The selection of oscillatory chemical processes for information processing applications comes from the fact that these processes are common and more robust than the acclaimed excitable systems. Chemical oscillations are observed at different spatio-temporal scales going down to nanostructures [39]. Therefore, methods that can help to simplify the design of an information processing medium based on oscillator networks can be important for their future success of chemical computing. It is important that information processing networks of oscillators designed with evolutionary algorithms are not based on the binary logic but use the properties of a medium in an optimized way. The algorithms are parallel just by their design.

As an example, I considered the application of optimization methods based on the maximization of mutual information between the classifier evolution and the record types of the test dataset for finding which of two numbers is larger. The optimized network was able to solve this problem with the accuracy exceeding 98%. Moreover, the evolutionary algorithm lead to a rational strategy that was utilizing the properties of the medium in a rational way. In order to speed-up the simulation procedure, I used the event based model of oscillations and interactions between oscillators, but there are no restrictions on the model. A similar functionality can be expected for oscillating reactions described by chemical kinetic equations. Therefore, I can claim that also in the other cases, the methods of information theory combined with evolutionary optimization are able to reveal the computing potential of the considered medium for the required computing task.

Funding: This research received no external funding.

Acknowledgments: A part of this work was supported by the Polish National Science Centre grant UMO-2014/15/B/ST4/04954.

Conflicts of Interest: The author declares no conflict of interest.

Abbreviations

The following abbreviations are used in this manuscript:

MDPI   Multidisciplinary Digital Publishing Institute
BZ   Belousov–Zhabotinsky

References

1. Feynman, R.P. *The Feynman Lectures on Computation*; Hey, A.J.G., Allen, R.W., Eds.; Addison-Wesley: Boston, MA, USA, 1996.
2. Haken, H. *Brain Dynamics, Synchronization and Activity Patterns in Pulse-Coupled Neural Nets with Delays and Noise*; Springer: Berlin/Heidelberg, Germany, 2002.
3. Toth, A.; Showalter, K. Logic gates in excitable media. *J. Chem. Phys.* **1995**, *103*, 2058–2066. [CrossRef]
4. Steinbock, O.; Kettunen, P.; Showalter, K. Chemical wave logic gates. *J. Phys. Chem.* **1996**, *100*, 18970–18975. [CrossRef]
5. Motoike, I.N.; Yoshikawa, K. Information operations with an excitable field. *Phys. Rev. E* **1999**, *59*, 5354–5360. [CrossRef]
6. Sielewiesiuk, J.; Gorecki, J. Logical functions of a cross junction of excitable chemical media. *J. Phys. Chem. A* **2001**, *105*, 8189–8195. [CrossRef]
7. Adamatzky, A.; De Lacy Costello, B. Experimental logical gates in a reaction–diffusion medium: the XOR gate and beyond. *Phys. Rev. E* **2002**, *66*, 046112. [CrossRef]
8. Steinbock, O.; Toth, A.; Showalter K. Navigating complex labyrinths—Optimal paths from chemical waves. *Science* **1995**, *267*, 868–871. [CrossRef]
9. Agladze, K; Magome, N.; Aliev, R.; Yamaguchi, T.; Yoshikawa, K. Finding the optimal path with the aid of chemical wave. *Physica D* **1997**, *106*, 247–254. [CrossRef]
10. Kuhnert, L. A new optical photochemical memory device in a light-sensitive chemical active medium. *Nature* **1986**, *319*, 393–394. [CrossRef]
11. Kuhnert, L.; Agladze, K.I.; Krinsky, V.I. Image processing using light-sensitive chemical waves. *Nature* **1989**, *337*, 244–247. [CrossRef]
12. Rambidi, N.G.; Maximychev, A.V. Towards a biomolecular computer. Information processing capabilities of biomolecular nonlinear dynamic media. *BioSystems* **1997**, *41*, 195–211. [CrossRef]
13. Tyson, J.J. What everyone should know about the Belousov–Zhabotinsky reaction. In *Frontiers in Mathematical Biology*; Levin, S.A., Ed.; Springer: Berlin/Heidelberg, Germany, 1994; pp. 569–587.
14. Epstein, I.R.; Pojman, J.A. *An Introduction to Nonlinear Chemical Dynamics: Oscillations, Waves, Patterns, and Chaos*; Oxford University Press: New York, NY, USA, 1998.
15. Sielewiesiuk, J.; Gorecki, J. Passive Barrier as a Transformer of Chemical Signal Frequency. *J. Phys. Chem. A* **2002**, *106*, 4068–4076. [CrossRef]
16. Adamatzky, A.; De Lacy Costello, B.; Asai, T. *Reaction–Diffusion Computers*; Elsevier: New York, NY, USA, 2005.
17. Burger, M.; Field, R.J. *Oscillations and Traveling Waves in Chemical Systems*; Wiley: New York, NY, USA, 1985.
18. Krug, H.-J.; Pohlmann, L.; Kuhnert, L. Analysis of the modified complete oregonator accounting for oxygen sensitivity and photosensitivity of Belousov-Zhabotinsky systems. *J. Phys. Chem.* **1990**, *94* , 4862–4866. [CrossRef]

19. Kádár, S.; Amemiya, T.; Showalter, K. Reaction mechanism for light sensitivity of the Ru(bpy)$_3^{2+}$-catalyzed Belousov-Zhabotinsky reaction. *J. Phys. Chem. A* **1997**, *101*, 8200–8206. [CrossRef]

20. Gizynski; K.; Gorecki, J. Chemical memory with states coded in light controlled oscillations of interacting Belousov–Zhabotinsky droplets. *Phys. Chem. Chem. Phys.* **2017**, *19*, 6519–6531. [CrossRef]

21. Gruenert, G.; Gizynski, K.; Escuela, G.; Ibrahim, B.; Gorecki, J.; Dittrich, P. Understanding Computing Droplet Networks by Following Information Flow. *Int. J. Neural Syst.* **2015**, *25*, 1450032. [CrossRef]

22. Gizynski, K.; Gruenert, G.; Dittrich, P.; Gorecki, J. Evolutionary design of classifiers made of droplets containing a nonlinear chemical medium. *MIT Evol. Comput.* **2017**, *25*, 643–671. [CrossRef]

23. Gizynski, K.; Gorecki, J. Cancer classification with a network of chemical oscillators. *Phys.Chem.* **2017**, *19*, 28808–28819. [CrossRef]

24. Fogel, D. B. An introduction to simulated evolutionary optimization. *IEEE Trans. Neural Netw.* **1994**, *5*, 3–14. [CrossRef]

25. Weicker, K. *Evolutionare Algorithmen*; Springer: Berlin/Heidelberg, Germany, 2007.

26. Gorecki, J.; Gorecka, J.N.; Adamatzky, A. Information coding with frequency of oscillations in Belousov–Zhabotinsky encapsulated disks. *Phys. Rev. E* **2014**, *89*, 042910. [CrossRef]

27. Gorecki, J.; Yoshikawa, K.; Igarashi, Y. On chemical reactors that can count. *J. Phys. Chem. A* **2003**, *107*, 1664–1669. [CrossRef]

28. Gizynski, K.; Gorecki, J. A Chemical System that Recognizes the Shape of a Sphere. *Comput. Methods Sci. Technol.* **2016**, *22*, 167–177. [CrossRef]

29. Muzika, F.; Schreiberova, L.; Schreiber, I. Chemical computing based on Turing patterns in two coupled cells with equal transport coefficients. *RSC Adv.* **2014**, *4*, 56165–56173. [CrossRef]

30. Muzika, F.; Schreiberova, L.; Schreiber, I. Discrete Turing patterns in coupled reaction cells in a cyclic array. *React. Kinet. Mech. Catal.* **2016**, *118*, 99–114. [CrossRef]

31. Szymanski, J.; Gorecka, J.N.; Igarashi, Y.; Gizynski, K.; Gorecki, J.; Zauner, K.P.; de Planque, M. Droplets with information processing ability. *Int. J. Unconv. Comput.* **2011**, *7*, 185–200.

32. Guzowski, J.; Gizynski, K.; Gorecki, J.; Garstecki, P. Microfluidic platform for reproducible self-assembly of chemically communicating droplet networks with predesigned number and type of the communicating compartments. *Lab Chip* **2016**, *16*, 764–772. [CrossRef]

33. Mallphanov, I. L.; Vanag, V.K. Fabrication of New Belousov-Zhabotinsky Micro-Oscillators on the Basis of Silica Gel Beads. *J. Phys. Chem. A* **2020**, *124*, 272–282. [CrossRef]

34. Kuze, M.; Horisaka, M.; Suematsu, N.J.; Amemiya, T.; Steinbock, O.; Nakata, S. Chemical Wave Propagation in the Belousov-Zhabotinsky Reaction Controlled by Electrical Potential. *J. Phys. Chem. A* **2019**, *123*, 4853–4857. [CrossRef]

35. Field, R.J.; Noyes, R.M. Oscillations in chemical systems. IV. Limit cycle behavior in a model of a real chemical reaction. *J. Chem. Phys.* **1974**, *60*, 1877–1884. [CrossRef]

36. Rovinsky, A.B.; Zhabotinsky, A.M.; Irving, R. Mechanism and mathematical model of the oscillating bromate–ferroin–bromomalonic acid reaction. *J. Phys. Chem.* **1984**, *88*, 6081–6084. [CrossRef]

37. Cover, T. M.; Thomas, J. A. *Elements of Information Theory*; Wiley-Interscience: New York, NY, USA, 2006.

38. Holley, J.; Jahan, I.; De Lacy Costello, B.; Bull, L.; Adamatzky, A. Logical and arithmetic circuits in Belousov-Zhabotinsky encapsulated disks. *Phys. Rev. E* **2011**, *84*, 056110. [CrossRef]

39. Grzybowski, B.A. *Chemistry in Motion: Reaction-Diffusion Systems for Micro- and Nanotechnology*; Wiley-Interscience: New York, NY, USA, 2009.

Article

Statistical Characteristics of Stationary Flow of Substance in a Network Channel Containing Arbitrary Number of Arms

Roumen Borisov [1], Zlatinka I. Dimitrova [2] and Nikolay K. Vitanov [1,3,*]

1   Institute of Mechanics, Bulgarian Academy of Sciences, Acad. G. Bonchev Str., Bl. 4, 1113 Sofia, Bulgaria; r.borisov@outlook.com
2   Georgi Nadjakov Institute of Solid State Physics, Bulgarian Academy of Sciences, Blvd. Tzarigradsko Chaussee 72, 1784 Sofia, Bulgaria; zdim@issp.bas.bg
3   Max-Planck Institute for the Physics of Complex Systems, Noethnitzerstr. 38, 01187 Dresden, Germany
*   Correspondence: vitanov@imbm.bas.bg

Received: 19 April 2020; Accepted: 12 May 2020; Published: 15 May 2020

Abstract: We study flow of substance in a channel of network which consists of nodes of network and edges which connect these nodes and form ways for motion of substance. The channel can have arbitrary number of arms and each arm can contain arbitrary number of nodes. The flow of substance is modeled by a system of ordinary differential equations. We discuss first a model for a channel which arms contain infinite number of nodes each. For stationary regime of motion of substance in such a channel we obtain probability distributions connected to distribution of substance in any of channel's arms and in entire channel. Obtained distributions are not discussed by other authors and can be connected to Waring distribution. Next, we discuss a model for flow of substance in a channel which arms contain finite number of nodes each. We obtain probability distributions connected to distribution of substance in the nodes of the channel for stationary regime of flow of substance. These distributions are also new and we calculate corresponding information measure and Shannon information measure for studied kind of flow of substance.

Keywords: network; flow; channel; probability distribution; Shannon information measure

---

1. Introduction

Complex systems have been studied intensively in the last decades as such systems are encountered frequently in the area natural sciences, population dynamics, social sciences, etc. [1–11]. Large number of phenomena in above systems can be studied by network models [12–16]. One class of phenomena is connected to certain kinds of motion of substance through channels of networks (such as, e.g., migration flows, logistic flows, transport of substances, etc.) [17–25]. In the last years we have studied several cases of flow of substance in channels of networks [23,26–31]. At the beginning, our studies have been inspired by application of studied models to channels of human migration. One direction of this research is related to probability distributions connected to motion of substance in studied channel of a network. At the beginning we have used model for channel with single arm. After that we have obtained results for channels containing two or three arms. Below we extend this theory for flow of substance in a class of channels containing arbitrary number of arms. We consider the situation where studied channel has a main arm which is the root for other arms. There are special nodes in this channel: the nodes where split of an arm happens. We shall consider the case where more than one arm can arise by this split.

The study presented below is carried out from point of view of possible application of discussed model to various practical cases. Because of this we use the general terms substance, channel and nodes of a network. The model can be applied to different kinds of substances and for flows of these substances in any channel which can be modeled by chains of nodes of a network. We note that the probability distributions obtained below are not discussed by other authors and these distributions can be connected to interesting long-tail distributions (e.g., to the Waring distribution) for particular case where the channel contains single arm.

The organization of text is as follows. In Section 2 we discuss a model of channel containing arms consisting of infinite number of nodes in each of them. We obtain probability distributions connected to amounts of substance in nodes of the arms of this channel for case of stationary flow of substance. Then we study the case of stationary flow of substance in channel which arms contain finite number of nodes each. In Section 3 we discuss the information measure and the Shannon information measure connected to the obtained probability distributions. Short discussion is presented in Section 4 and several concluding remarks are summarized in Section 5.

2. Results

2.1. Flow of Substance in a Channel Consisting of Arms Containing Infinite Number of Nodes Each

The model of motion of substance through the channel discussed below is an extension of model discussed in [23,24]. The channel consists of chains of nodes of a network—Figure 1.

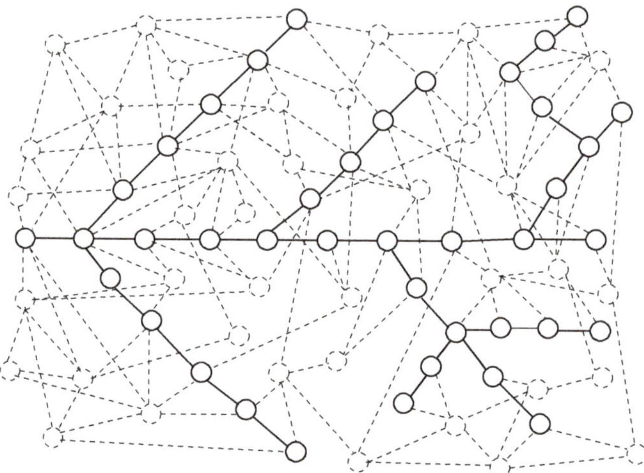

Figure 1. Network and studied channel. Nodes and edges which belong to the channel are marked by solid lines. Other nodes and edges of the network are marked by dashed lines.

The convention for numbering of nodes of channel is as follows—Figure 2. Let us denote a node of the channel by $\mathcal{N}$. We associate 4 indexes to each node: $\mathcal{N}_{i,j}^{a,b}$. The lower indexes specify position of node in current arm. i is the number associated with current arm. j is the the number of node of i-th arm. Upper indexes specify the origin of arm i. The index a is number of arm from which arm i splits. The index b is number of node of arm a where this split happens. Thus $a = 3, b = 8$ means: arm i of the channel arises from node 8 of arm 3. Then $\mathcal{N}_{3,5}^{1,2}$ means: 5-th node of arm 3 which splits at node 2 of channel's arm 1.

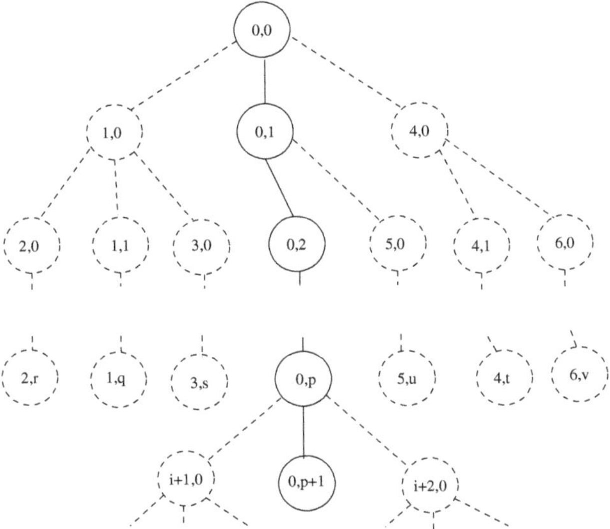

Figure 2. The channel and numbering of its nodes. Only the lower two indexes of numbering of nodes are shown.

We assume that some substance can enter studied channel from external environment only through the 0-th node of main arm of the channel (this arm is labeled by $q = 0$ below in the text). In addition the substance can move only in one direction in any of arms (from nodes labeled by smaller values of index j to nodes labeled by larger values of index j). Nodes of each arm are connected by edges and each node is connected only to two neighboring nodes of the arm except for special nodes where a split of an arm happens. These special nodes can be connected to 1 or more additional nodes. In addition we assume that substance can quit the nodes of channel and can move to environment. This process will be called "leakage". As substance can enter the channel only through 0-th node of main arm then leakage is possible only in direction from channel nodes to other network nodes (and not in the opposite direction).

We stress the following. The node where arm i begins is labeled as 0-th node of i-th arm. This node is the next one after splitting at node b of arm a. Thus any of nodes of the channel has unique notation. This is illustrated in Figure 2. The 0-th node of arm 1 arises from 0-th node of the arm 0 (the 0-th node of arm 0 is "environment" which supplies substance to 0-th node of arm 1).

We can consider each node as a cell (box), in other words, we consider the arm to be an array of infinite number of cells indexed in succession by non-negative integers. We assume that an amount $x_q^{a_q,b_q}$ of some substance is distributed among cells of the arm q which splits at node (a_q, b_q) of the network. This substance can move from one cell to another cell.

Let $x_{q,i}^{a_q,b_q}$ be amount of substance in i-th cell of the q-th arm of the channel. We consider in this section a channel containing infinite number of nodes in each of its arms. Then

$$x_q^{a_q,b_q} = \sum_{i=0}^{\infty} x_{q,i}^{a_q,b_q}. \tag{1}$$

The fractions $y_{q,i}^{a_q,b_q} = x_{q,i}^{a_q,b_q} / x_q^{a_q,b_q}$ can be considered as probability values of distribution of a discrete random variable ζ_q in corresponding arm of channel

$$y_{q,i}^{a_q,b_q} = p_q^{a_q,b_q}(\zeta_q = i), \ i = 0,1,\ldots \tag{2}$$

We can define another distribution: the distribution of substance in entire channel. The total amount of substance in this case is

$$x = \sum_{q=0}^{M} x_q^{a_q,b_q} = \sum_{q=0}^{M} \sum_{i=0}^{\infty} x_{q,i}^{a_q,b_q},$$ (3)

where $M+1$ is number of arms of channel (remember the main arm of channel which has number 0). Corresponding distribution of substance in the channel is

$$z_{q,i}^{a_q,b_q} = \frac{x_{q,i}^{a_q,b_q}}{x}$$ (4)

Next we assume that the amount $x_{q,i}^{a_q,b_q}$ of substance in i-th node of q-th arm of channel can change because of following processes:

1. Some amount $s_q^{a,b}$ of substance enters arm q from external environment through 0-th cell of corresponding arm. We consider two kinds of external environments for an arm of the channel

   (a) For the root of the channel (arm with label $q = 0$): substance $s_0^{0,0}$ enters the root through environment of the channel
   (b) For the arms of the channel which are not root (i.e., which number is $q \neq 0$): Substance $s_q^{a,b}$ is part of substance presented in node (a, b) of parent arm. This substance "leaks" from the parent arm to corresponding child arm.

   The substance $s_q^{a,b}$ is presented only in node 0 of q-th arm of channel. For other nodes of channel there is no substance which enters the node from environment of channel.

2. Amount $f_{q,i}^{a,b}$ from $x_{q,i}^{a,b}$ is transferred from the i-th cell to $(i+1)$-th cell of q-th arm;

3. Amount $g_{q,i}^{a,b}$ of $x_{q,i}^{a,b}$ leaks out i-th cell of q-th arm to environment of the arm of channel. This leakage can be of two kinds

   (a) Leakage to the environment of channel: this kind of leakage leads to loss of substance for the channel
   (b) Leakage to other arms of the channel which begin from the node b of the arm a: This leakage is connected to the substance $s_q^{a,b}$ which enters corresponding child arm of channel which splits from node b of arm a.

We assume that the process of motion of substance is continuous in time. Then the motion of substance among nodes of q-th arm can be modeled mathematically by a system of ordinary differential equations:

$$\frac{dx_{q,0}^{a,b}}{dt} = s_q^{a,b} - f_{q,0}^{a,b} - g_{q,0}^{a,b};$$

$$\frac{dx_{i,q}^{a,b}}{dt} = f_{q,i-1}^{a,b} - f_{q,i}^{a,b} - g_{q,i}^{a,b}, \ i = 1, 2, \ldots,;$$ (5)

There are two regimes of work of the channel: stationary regime and non-stationary regime. We shall discuss below the stationary regime of work. In this regime $dx_{i,q}^{a,b}/dt = 0, i = 0, 1, \ldots$. Let us mark the quantities for the stationary regime with *. Then from (5) one obtains

$$f_{q,0}^{*a,b} = s_q^{*a,b} - g_{q,0}^{*a,b}; \ f_{q,i}^{*a,b} = f_{q,i-1}^{*a,b} - g_{q,i}^{*a,b}$$ (6)

We assume the following relationships for amounts of moving substances in (5) ($\alpha_i, \beta_i, \gamma_i, \sigma_0$ are parameters):

$$
\begin{aligned}
s_0^{0,0} &= \sigma_0 x_{0,0}^{0,0} > 0 \\
s_q^{a,b} &= \delta_q x_{a,b}^{c,d}; \ 1 \geq \delta_q \geq 0 \\
f_{q,i}^{a,b} &= (\alpha_{q,i}^{a,b} + i\beta_{q,i}^{a,b})x_{q,i}^{a,b}; \ 1 > \alpha_{q,i}^{a,b} > 0, \ 1 \geq \beta_{q,i}^{a,b} \geq 0 \\
g_{q,i}^{a,b} &= \hat{\gamma}_{q,i}^{a,b} x_{q,i}^{a,b}; \ 1 \geq \hat{\gamma}_{q,i}^{a,b} \geq 0 \rightarrow \text{non-uniform} \\
&\qquad \text{leakage in the nodes.}
\end{aligned}
\tag{7}
$$

Indexes c and d in the second of above relationships describe parent arm (numbered by c) and parent node of the arm c (numbered by d) for the arm q. $\beta_{q,i}^{a,b}$ accounts for circumstances which lead substance to leave faster the node i. $\hat{\gamma}_{q,i}^{a,b}$ is a quantity specific for the present study. $\hat{\gamma}_{q,i}^{a,b} = \gamma_{q,i}^{a,b} + \sum_{p\in(q,i)} \delta_{p,q,i}^{a,b}$ describes the situation with leakages in cells. $\gamma_{q,i}^{a,b}$ is the leakage to environment from i-th node of q-th arm. $\delta_{p,q,i}^{a,b}$ describes the leakage to the nodes which split from i-th node of q-th arm. The notation $p \in (q,i)$ in the sum means all arms which arise from node i of arm q.

On the basis of all above the model system of differential equations for q-th arm of channel becomes

$$
\begin{aligned}
\frac{dx_{q,0}^{a,b}}{dt} &= s_q^{a,b} - \alpha_{q,0}^{a,b} x_{q,0}^{a,b} - \hat{\gamma}_{q,0}^{a,b} x_{q,0}^{a,b}; \\
\frac{dx_{q,i}^{a,b}}{dt} &= [\alpha_{q,i-1}^{a,b} + (i-1)\beta_{q,i-1}^{a,b}]x_{q,i-1}^{a,b} - (\alpha_{q,i}^{a,b} + i\beta_{q,i}^{a,b} + \hat{\gamma}_{q,i}^{a,b})x_{q,i}^{a,b}; \ i = 1,2,\ldots
\end{aligned}
\tag{8}
$$

Below we shall discuss the situation in which a stationary state exists in entire channel. Then we have $dx_0^q/dt = 0$ in first of the Equations (8). Hence

$$
x_{q,0}^{*a,b} = \frac{s_q^{a,b}}{\alpha_{0,q}^{a,b} + \hat{\gamma}_{0,q}^{a,b}} \ .
\tag{9}
$$

For the root of the channel (arm 0) we substitute $s_0^{0,0}$ from (7) in (9) and obtain that $x_{0,0}^{0,0}$ is a free parameter and in addition

$$
\sigma_0 = \alpha_{0,0}^{0,0} + \hat{\gamma}_{0,0}^{0,0}.
$$

For the arm r which arises from node m of arm q, $dx_{r,0}^{q,m}/dt = 0$ and then from model equations above we obtain

$$
x_{r,0}^{*q,m} = \frac{\delta_{r,q,m}^{a,b}}{\alpha_{r,0}^{q,m} + \hat{\gamma}_{r,0}^{q,m}} x_{q,m}^{*c,d}.
\tag{10}
$$

In principle the solution of Equations (8) is

$$
\begin{aligned}
x_{q,i}^{a,b} &= x_{q,i}^{*a,b} + \sum_{j=0}^{i} b_{q,i,j}^{a,b} \exp[-(\alpha_{q,j}^{a,b} + j\beta_{q,j}^{a,b} + \hat{\gamma}_{q,j}^{a,b})t], \\
&\quad i = 1,2,\ldots,
\end{aligned}
\tag{11}
$$

where $x_{q,i}^{*a,b}$ is stationary part of solution. We note that because of the non-negative values of the parameters α, β, and $\hat{\gamma}$, $x_{q,i}^{a,b}$ converges to $x_{q,i}^{*a,b}$ with increasing time.

For $x_{q,i}^{*a,b}$ one obtains the relationship (just set $dx_{q,i}^{a,b}/dt = 0$ in Equations (8))

$$x_{q,i}^{*a,b} = \frac{\alpha_{q,i-1}^{a,b} + (i-1)\beta_{q,i-1}^{a,b}}{\alpha_{q,i}^{a,b} + i\beta_{q,i}^{a,b} + \hat{\gamma}_{q,i}^{a,b}} x_{q,i-1}^{*a,b}, \quad i = 1, 2, \dots. \tag{12}$$

The corresponding relationships for coefficients b_{ij}^{q} are ($i = 1, \dots$):

$$b_{q,i,j}^{a,b} = \frac{\alpha_{q,i-1}^{a,b} + (i-1)\beta_{q,i-1}^{a,b}}{(\alpha_{q,i}^{a,b} - \alpha_{q,i}^{a,b}) + (i\beta_{q,i}^{a,b} - j\beta_{q,j}^{a,b}) + (\hat{\gamma}_{q,i}^{a,b} - \hat{\gamma}_{q,j}^{a,b})} b_{q,i-1,j}^{a,b},$$
$$j = 0, 1, \dots, i-1, \tag{13}$$

From Equation (12) one obtains ($i = 1, 2, \dots$)

$$x_{q,i}^{*a,b} = \frac{\prod\limits_{j=0}^{i-1}\left[\alpha_{q,i-j-1}^{a,b} + (i-j-1)\beta_{q,i-j-1}^{a,b}\right]}{\prod\limits_{j=0}^{i-1}\left[\alpha_{q,i-j}^{a,b} + (i-j)\beta_{q,i-j}^{a,b} + \hat{\gamma}_{q,i-j}^{a,b}\right]} x_{q,0}^{*a,b} \tag{14}$$

The total amount of substance in the nodes of arm q is

$$x_q^{*a,b} = \sum_{i=0}^{\infty} x_{q,i}^{*a,b} = x_{q,0}^{*a,b} + \sum_{i=1}^{\infty} \frac{\prod\limits_{j=0}^{i-1}\left[\alpha_{q,i-j-1}^{a,b} + (i-j-1)\beta_{q,i-j-1}^{a,b}\right]}{\prod\limits_{j=0}^{i-1}\left[\alpha_{q,i-j}^{a,b} + (i-j)\beta_{q,i-j}^{a,b} + \hat{\gamma}_{q,i-j}^{a,b}\right]} x_{q,0}^{*a,b}. \tag{15}$$

The form of corresponding probability distribution $y_{q,i}^{*a,b} = x_{q,i}^{*a,b} / x_q^{*a,b}$ is

$$y_{q,0}^{*a,b} = \frac{1}{1 + \sum\limits_{i=1}^{\infty} \dfrac{\prod\limits_{j=0}^{i-1}\left[\alpha_{q,i-j-1}^{a,b} + (i-j-1)\beta_{q,i-j-1}^{a,b}\right]}{\prod\limits_{j=0}^{i-1}\left[\alpha_{q,i-j}^{a,b} + (i-j)\beta_{q,i-j}^{a,b} + \gamma_{q,i-j}^{*a,b}\right]}}$$

$$y_{q,i}^{*a,b} = \frac{\dfrac{\prod\limits_{j=0}^{i-1}\left[\alpha_{q,i-j-1}^{a,b} + (i-j-1)\beta_{q,i-j-1}^{a,b}\right]}{\prod\limits_{j=0}^{i-1}\left[\alpha_{q,i-j}^{a,b} + (i-j)\beta_{q,i-j}^{a,b} + \hat{\gamma}_{q,i-j}^{a,b}\right]}}{1 + \sum\limits_{i=1}^{\infty} \dfrac{\prod\limits_{j=0}^{i-1}\left[\alpha_{q,i-j-1}^{a,b} + (i-j-1)\beta_{q,i-j-1}^{a,b}\right]}{\prod\limits_{j=0}^{i-1}\left[\alpha_{q,i-j}^{a,b} + (i-j)\beta_{q,i-j}^{a,b} + \hat{\gamma}_{q,i-j}^{a,b}\right]}}$$
$$i = 1, 2, \dots \tag{16}$$

We can write probability distribution connected to distribution of substance in a channel containing M arms ($M = 1, 2, \dots$). The total amount of substance in the arms of the channel is

$$x^* = \sum_{q=0}^{M} x_{q,0}^{*a,b}\left[1 + \sum_{i=1}^{\infty} \frac{\prod\limits_{j=0}^{i-1}\left[\alpha_{q,i-j-1}^{a,b} + (i-j-1)\beta_{q,i-j-1}^{a,b}\right]}{\prod\limits_{j=0}^{i-1}\left[\alpha_{q,i-j}^{a,b} + (i-j)\beta_{q,i-j}^{a,b} + \hat{\gamma}_{q,i-j}^{a,b}\right]}\right]. \tag{17}$$

The probability distribution connected to entire channel is as follows. For the 0-th node of p-th arm of the channel

$$z_{p,0}^{*a_p,b_p} = \frac{x_{p,0}^{*,a_p,b_p}}{\sum\limits_{q=0}^{M} x_{q,0}^{*a,b} \left[1 + \sum\limits_{i=1}^{\infty} \frac{\prod\limits_{j=0}^{i-1} \left[\alpha_{q,i-j-1}^{a,b} + (i-j-1)\beta_{q,i-j-1}^{a,b} \right]}{\prod\limits_{j=0}^{i-1} \left[\alpha_{q,i-j}^{a,b} + (i-j)\beta_{q,i-j}^{a,b} + \hat{\gamma}_{q,i-j}^{a,b} \right]} \right]}, \tag{18}$$

and for the i-th node of the p-th arm of the channel ($i = 1, 2,,\ldots$)

$$z_{p,i}^{*a_p,b_p} = \frac{\dfrac{\prod\limits_{j=0}^{i-1} \left[\alpha_{p,i-j-1}^{a,b} + (i-j-1)\beta_{p,i-j-1}^{a,b} \right]}{\prod\limits_{j=0}^{i-1} \left[\alpha_{p,i-j}^{a,b} + (i-j)\beta_{p,i-j}^{a,b} + \hat{\gamma}_{p,i-j}^{a,b} \right]} x_{p,0}^{*a,b}}{\sum\limits_{q=0}^{M} x_{q,0}^{*a,b} \left[1 + \sum\limits_{i=1}^{\infty} \frac{\prod\limits_{j=0}^{i-1} \left[\alpha_{q,i-j-1}^{a,b} + (i-j-1)\beta_{q,i-j-1}^{a,b} \right]}{\prod\limits_{j=0}^{i-1} \left[\alpha_{q,i-j}^{a,b} + (i-j)\beta_{q,i-j}^{a,b} + \hat{\gamma}_{q,i-j}^{a,b} \right]} \right]} \tag{19}$$

To the best of our knowledge the distributions presented by (16), (18) and (19) have been not discussed up to now outside our research group, in other words, they are new probability distributions. We note that for case of channel containing just one arm the obtained probability distributions reduce to distribution discussed in the Appendix of [23]. This distribution is connected to the long-tail distribution of Waring (Edward Waring was the 6-th Lucasian professor in mathematics at University of Cambridge).

2.2. Theory for the Case of Channel Consisting of Arms Containing Finite Number of Nodes

We consider a channel containing main arm labeled by 0 and number M of other arms. The arm q of this channel has finite number of $N_q + 1$ nodes (labeled from 0 to N_q). The mathematical model for this case consists of a system of equations which contains an equation for 0-th node, equations for nodes $1, \ldots, N_{q-1}$ and equation for node N_q. The model system of equations for node 0 and for nodes $1, \ldots, N_{q-1}$ of q-th arm of the channel is (notations are the same as in the previous section)

$$\frac{dx_{q,0}^{a,b}}{dt} = s_q^{a,b} - \alpha_{q,0}^{a,b} x_{q,0}^{a,b} - \hat{\gamma}_{q,0}^{a,b} x_{q,0}^{a,b}, \tag{20}$$

$$\frac{dx_{q,i}^{a,b}}{dt} = [\alpha_{q,i-1}^{a,b} + (i-1)\beta_{q,i-1}^{a,b}] x_{q,i-1}^{a,b} - (\alpha_{q,i}^{a,b} + i\beta_{q,i}^{a,b} + \hat{\gamma}_{q,i}^{a,b}) x_{q,i}^{a,b};$$
$$i = 1, 2, \ldots, N_q - 1. \tag{21}$$

For node N_q of q-th arm there is no outflow to next mode of the arm (as node N_q is the last node of q-th arm). Thus the equation for motion of substance for this node is

$$\frac{dx_{q,N}^{a,b}}{dt} = [\alpha_{q,N_q-1}^{a,b} + (N_q - 1)\beta_{q,N_q-1}^{a,b}] x_{q,N_q-1}^{a,b} - \hat{\gamma}_{q,N_q}^{a,b} x_{q,N_q}^{a,b}. \tag{22}$$

We discuss the case of stationary motion of substance through arms of studied channel. Then $dx_0^q/dt = 0$ in (20). Hence

$$x_{q,0}^{*a,b} = \frac{s_q^{a,b}}{\alpha_{0,q}^{a,b} + \hat{\gamma}_{0,q}^{a,b}}. \tag{23}$$

For root of the channel (arm 0) we substitute $s_0^{0,0}$ from Equation (7) in Equation (23) and obtain that $x_{0,0}^{0,0}$ is a free parameter and in addition

$$\sigma_0 = \alpha_{0,0}^{0,0} + \hat{\gamma}_{0,0}^{0,0}.$$

For arm r which arises from node m of arm q, $dx_{r,0}^{q,m}/dt = 0$ and then from the model equations above we obtain

$$x_{r,0}^{*q,m} = \frac{\delta_{r,q,m}^{a,b}}{\alpha_{r,0}^{q,m} + \hat{\gamma}_{r,0}^{q,m}} x_{q,m}^{*c,d}. \tag{24}$$

For $x_{q,i}^{*a,b}$ we obtain the relationship (just set $dx_{q,i}^{a,b}/dt = 0$ in (21))

$$x_{q,i}^{*a,b} = \frac{\alpha_{q,i-1}^{a,b} + (i-1)\beta_{q,i-1}^{a,b}}{\alpha_{q,i}^{a,b} + i\beta_{q,i}^{a,b} + \hat{\gamma}_{q,i}^{a,b}} x_{q,i-1}^{*a,b}, \quad i = 1, 2, \ldots, N_q - 1; \tag{25}$$

In order to calculate $x_{q,N_q}^{*a,b}$ we use (22). The result is

$$x_{q,N_q}^{*a,b} = \frac{\alpha_{q,N_q-1}^{a,b} + (N_q - 1)\beta_{q,N_q-1}^{a,b}}{\hat{\gamma}_{q,N_q}^{a,b}} x_{q,N_q-1}^{*a,b}. \tag{26}$$

What follows from (25) is

$$x_{q,i}^{*a,b} = \frac{\prod_{j=0}^{i-1}\left[\alpha_{q,i-j-1}^{a,b} + (i-j-1)\beta_{q,i-j-1}^{a,b}\right]}{\prod_{j=0}^{i-1}\left[\alpha_{q,i-j}^{a,b} + (i-j)\beta_{q,i-j}^{a,b} + \hat{\gamma}_{q,i-j}^{a,b}\right]} x_{q,0}^{*a,b}, \quad i = 1, \ldots, N_q - 1. \tag{27}$$

And from (26) we obtain

$$x_{q,N_q}^{*a,b} = x_{q,0}^{*a,b} \frac{\alpha_{q,N_q-1}^{a,b} + (N_q - 1)\beta_{q,N_q-1}^{a,b}}{\hat{\gamma}_{q,N_q}^{a,b}} \times$$

$$\frac{\prod_{j=0}^{N_q-2}\left[\alpha_{q,N_q-j-2}^{a,b} + (N_q - j - 2)\beta_{q,N_q-j-2}^{a,b}\right]}{\prod_{j=0}^{N_q-2}\left[\alpha_{q,N_q-j-1}^{a,b} + (N_q - j - 1)\beta_{q,N_q-j-1}^{a,b} + \hat{\gamma}_{q,N_q-j-1}^{a,b}\right]}. \tag{28}$$

The total amount of the substance in q-th arm of channel is

$$x_q^{*,a,b} = x_{q,0}^{*a,b} + \sum_{i=1}^{N_q-1} x_{q,i}^{*,a,b} + x_{q,N}^{*,a,b} = x_{q,0}^{*a,b} A_q \tag{29}$$

where A is given by relationship

$$A_q = 1 + \sum_{i=1}^{N_q-1} \frac{\prod_{j=0}^{i-1}\left[\alpha_{q,i-j-1}^{a,b} + (i-j-1)\beta_{q,i-j-1}^{a,b}\right]}{\prod_{j=0}^{i-1}\left[\alpha_{q,i-j}^{a,b} + (i-j)\beta_{q,i-j}^{a,b} + \hat{\gamma}_{q,i-j}^{a,b}\right]} + \frac{\alpha_{q,N_q-1}^{a,b} + (N_q-1)\beta_{q,N_q-1}^{a,b}}{\hat{\gamma}_{q,N_q}^{a,b}} \times$$
$$\frac{\prod_{j=0}^{N_q-2}\left[\alpha_{q,N_q-j-2}^{a,b} + (N_q-j-2)\beta_{q,N_q-j-2}^{a,b}\right]}{\prod_{j=0}^{N_q-2}\left[\alpha_{q,N_q-j-1}^{a,b} + (N_q-j-1)\beta_{q,N_q-j-1}^{a,b} + \hat{\gamma}_{q,N_q-j-1}^{a,b}\right]}. \tag{30}$$

The distribution of substance in nodes of q-th arm of channel is

$$y_{q,0}^{*a,b} = \frac{1}{A_q}, \tag{31}$$

$$y_{q,i}^{*a,b} = \frac{B_{q,i}}{A_q}, \tag{32}$$

where

$$B_{q,i} = \frac{\prod_{j=0}^{i-1}\left[\alpha_{q,i-j-1}^{a,b} + (i-j-1)\beta_{q,i-j-1}^{a,b}\right]}{\prod_{j=0}^{i-1}\left[\alpha_{q,i-j}^{a,b} + (i-j)\beta_{q,i-j}^{a,b} + \hat{\gamma}_{q,i-j}^{a,b}\right]} \quad i = 1,\dots,N_q-1, \tag{33}$$

$$y_{q,N_q}^{*a,b} = \frac{B_{q,N_q}}{A_q}, \tag{34}$$

where B_{q,N_q} is given by the relationship

$$B_{q,N_q} = \frac{\alpha_{q,N_q-1}^{a,b} + (N_q-1)\beta_{q,N_q-1}^{a,b}}{\hat{\gamma}_{q,N_q}^{a,b}} \times$$
$$\frac{\prod_{j=0}^{N_q-2}\left[\alpha_{q,N_q-j-2}^{a,b} + (N_q-j-2)\beta_{q,N_q-j-2}^{a,b}\right]}{\prod_{j=0}^{N_q-2}\left[\alpha_{q,N_q-j-1}^{a,b} + (N_q-j-1)\beta_{q,N_q-j-1}^{a,b} + \hat{\gamma}_{q,N_q-j-1}^{a,b}\right]}. \tag{35}$$

We can write also the probability distribution of substance for entire channel, in other words, for M branches of channel. The total amount of substance in this case is

$$x^* = \sum_{q=0}^{M} x_{q,0}^{*a_q,b_q} A_q. \tag{36}$$

Distribution of substance in entry nodes of arms ($p = 0,\dots,M$) of the channel is:

$$z_{p,0}^{*a_p,b_p} = \frac{x_{p,0}^{*a_p,b_p}}{\sum_{q=0}^{M} x_{q,0}^{*a_q,b_q} A_q}. \tag{37}$$

Distribution of substance in interior nodes of the arms is:

$$z_{p,i}^{*a_p,b_p} = \frac{x_{p,i}^{*,a_p,b_p}}{\sum\limits_{q=0}^{M} x_{q,0}^{*a_q,b_q} A_q} = \frac{B_{p,i} x_{p,0}^{*a_p,b_p}}{\sum\limits_{q=0}^{M} x_{q,0}^{*a_q,b_q} A_q}, \quad i = 1, \ldots, N_{q-1}. \tag{38}$$

Distribution of substance in last nodes of arms of the channel is:

$$z_{p,N_p}^{*a_p,b_p} = \frac{x_{p,N_p}^{*a_p,b_p}}{\sum\limits_{q=0}^{M} x_{q,0}^{*a_q,b_q} A_q} = \frac{B_{p,N_p} x_{p,0}^{*a_p,b_p}}{\sum\limits_{q=0}^{M} x_{q,0}^{*a_q,b_q} A_q}. \tag{39}$$

To the best of our knowledge the distributions presented by (31)–(34) and (37)–(39) are not discussed up to now outside our research group. In other words, these are new probability distributions. The obtained distributions are interesting for the practice as they are connected to class of channels containing finite number of arms and in addition each arm contains finite number of nodes.

3. Information Measures Connected to Obtained Probability Distributions

We can calculate various quantities connected to the obtained distributions. Below we consider an example related to an information problem. Let us consider flow of substance in the channel of network studied above. Each node of the channel is numbered and we can consider these nodes as letters of an alphabet. Let some kind of event happens in any of the nodes and let probability of occurrence of this event be proportional of amount of substance in corresponding node. Thus probability of occurrence of event in a node of channel will be equal to the probability from the corresponding probability distribution obtained above in the text. The channel (the source) will generate events with corresponding probability and we can calculate measure of information and Shannon measure of information for these sequences.

The information measure connected to an event with probability p is

$$I_p = -\log(p), \tag{40}$$

and Shannon information measure (average information we get from a symbol in a stream) connected to probability distribution $P = (p_0, \ldots, p_N)$ is

$$H(P) = -\sum_{i=0}^{N} p_i \log(p_i) \tag{41}$$

Let us consider the distribution P^* of substance in q-th arm of the channel given by (31)–(35). The information connected to event with probability p_i from i-th node of this arm is

$$\begin{aligned}
I_{p_0} &= \log(A_q) \\
I_{p_i} &= \log(A_q) - \log(B_{q,i}), \quad i = 1, \ldots, N_q - 1 \\
I_{p_N} &= \log(A_q) - \log(B_{q,N_q}).
\end{aligned} \tag{42}$$

The Shannon information measure connected to distribution P^* is

$$H(P^*) = \frac{\log(A_q)}{A_q} - \sum_{i=1}^{N_q-1} \frac{B_{q,i}}{A_q} \log\left(\frac{B_{q,i}}{A_q}\right) - \frac{B_{q,N_q}}{A_q} \log\left(\frac{B_{q,N_q}}{A_q}\right). \tag{43}$$

Let us consider now a very simple case: a channel containing single arm that has just three nodes. Below we write information measures for the nodes as well as Shannon information measure for this channel. We shall omit the indices a, b and q. We note that for this case $N_q = N = 2$ as labels of nodes of the arm are 0, 1 and 2. The probabilities connected to three nodes of studied channel are

$$
\begin{aligned}
p_0 &= \frac{1}{1 + \frac{\alpha_0}{\alpha_1 + \beta_1 + \hat{\gamma}_1}\left(1 + \frac{\alpha_1 + \beta_1}{\hat{\gamma}_2}\right)}, \\[2ex]
p_1 &= \frac{\frac{\alpha_0}{\alpha_1 + \beta_1 + \hat{\gamma}_1}}{1 + \frac{\alpha_0}{\alpha_1 + \beta_1 + \hat{\gamma}_1}\left(1 + \frac{\alpha_1 + \beta_1}{\hat{\gamma}_2}\right)}, \\[2ex]
p_2 &= \frac{\frac{\alpha_1 + \beta_1}{\hat{\gamma}_2}\frac{\alpha_0}{\alpha_1 + \beta_1 + \hat{\gamma}_1}}{1 + \frac{\alpha_0}{\alpha_1 + \beta_1 + \hat{\gamma}_1}\left(1 + \frac{\alpha_1 + \beta_1}{\hat{\gamma}_2}\right)}.
\end{aligned}
\tag{44}
$$

The parameters in (44) account for following processes

- α_0 $(0 < \alpha_0 < 1)$: flow between first and second node,
- α_1 $(0 < \alpha_1 < 1)$: flow between second and third node,
- β_1 $(0 < \beta_1 < 1 - \alpha_1)$: preference for the third node,
- $\hat{\gamma}_1$ $(0 \le \hat{\gamma}_1 < 1 - \alpha_1 - \beta_1)$: leakage from the second node,
- $\hat{\gamma}_2$ $(0 < \hat{\gamma}_2 \le 1)$: leakage from the third node.

The information measures connected to nodes are

$$
\begin{aligned}
I_{p_0} &= \log\left[1 + \frac{\alpha_0}{\alpha_1 + \beta_1 + \hat{\gamma}_1}\left(1 + \frac{\alpha_1 + \beta_1}{\hat{\gamma}_2}\right)\right] \\[2ex]
I_{p_1} &= \log\left[1 + \frac{\alpha_0}{\alpha_1 + \beta_1 + \hat{\gamma}_1}\left(1 + \frac{\alpha_1 + \beta_1}{\hat{\gamma}_2}\right)\right] - \log\left(\frac{\alpha_0}{\alpha_1 + \beta_1 + \hat{\gamma}_1}\right) \\[2ex]
I_{p_2} &= \log\left[1 + \frac{\alpha_0}{\alpha_1 + \beta_1 + \hat{\gamma}_1}\left(1 + \frac{\alpha_1 + \beta_1}{\hat{\gamma}_2}\right)\right] - \\[1ex]
&\quad \log\left(\frac{\alpha_1 + \beta_1}{\hat{\gamma}_2}\frac{\alpha_0}{\alpha_1 + \beta_1 + \hat{\gamma}_1}\right).
\end{aligned}
\tag{45}
$$

The corresponding Shannon information measure is

$$
\begin{aligned}
H &= \frac{1}{\left[1 + \frac{\alpha_0}{\alpha_1 + \beta_1 + \hat{\gamma}_1}\left(1 + \frac{\alpha_1 + \beta_1}{\hat{\gamma}_2}\right)\right]}\log\left[1 + \frac{\alpha_0}{\alpha_1 + \beta_1 + \hat{\gamma}_1}\left(1 + \frac{\alpha_1 + \beta_1}{\hat{\gamma}_2}\right)\right] + \\[2ex]
&\quad \frac{\left(\frac{\alpha_0}{\alpha_1 + \beta_1 + \hat{\gamma}_1}\right)}{\left[1 + \frac{\alpha_0}{\alpha_1 + \beta_1 + \hat{\gamma}_1}\left(1 + \frac{\alpha_1 + \beta_1}{\hat{\gamma}_2}\right)\right]}\left\{\log\left[1 + \frac{\alpha_0}{\alpha_1 + \beta_1 + \hat{\gamma}_1}\left(1 + \frac{\alpha_1 + \beta_1}{\hat{\gamma}_2}\right)\right] - \right. \\[2ex]
&\quad \left. \log\left(\frac{\alpha_0}{\alpha_1 + \beta_1 + \hat{\gamma}_1}\right)\right\} + \frac{\left(\frac{\alpha_1 + \beta_1}{\hat{\gamma}_2}\frac{\alpha_0}{\alpha_1 + \beta_1 + \hat{\gamma}_1}\right)}{\left[1 + \frac{\alpha_0}{\alpha_1 + \beta_1 + \hat{\gamma}_1}\left(1 + \frac{\alpha_1 + \beta_1}{\hat{\gamma}_2}\right)\right]} \times \\[2ex]
&\quad \left\{\log\left[1 + \frac{\alpha_0}{\alpha_1 + \beta_1 + \hat{\gamma}_1}\left(1 + \frac{\alpha_1 + \beta_1}{\hat{\gamma}_2}\right)\right] - \log\left(\frac{\alpha_1 + \beta_1}{\hat{\gamma}_2}\frac{\alpha_0}{\alpha_1 + \beta_1 + \hat{\gamma}_1}\right)\right\}.
\end{aligned}
\tag{46}
$$

Several illustrations for the dependence of $p_{0,1,2}$, I_{p_0, p_1, p_2} and H on parameters of problem are presented in Figures 3–7.

Figure 3 shows influence of the parameters of problem on probabilities $p_{0,1,2}$ (connected to stationary distribution of substance in the three nodes of studied channel). Figure 3a shows the influence of α_0 on $p_{0,1,2}$ when other parameters are fixed. α_0 is a parameter which regulates the outflow of substance from node 0 to node 1 of channel. For this case the probabilities can be written as follows:

$$
p_0 = \frac{1}{1 + c_0\alpha_0}; \quad p_1 = \frac{c_1\alpha_0}{1 + c_0\alpha_0}; \quad p_2 = \frac{c_2\alpha_0}{1 + c_0\alpha_0},
$$

where c_0, c_1, c_2 are appropriate constants which values depend on values of fixed parameters.

With increasing outflow from the node 0, p_0 decreases and p_1 and p_2 increase. This is connected to redistribution of percentage of total substance which is presented in any of the three nodes: percentage of the substance in node 0 decreases because of the increased outflow and this leads to increase of percentage of substance in other two nodes.

Figure 3b shows influence of increasing value of parameter α_1 on the probabilities $p_{0,1,2}$. Parameter α_1 accounts for the outflow of substance from node 1 to node 2 of studied channel. Increase of the value of α_1 leads to decrease of percentage of substance in node 2 and to increase of percentage of substance in node 2. Interesting is what happens in node 0. For fixed values of parameters as in Figure 3b the percentage of substance in node 0 increases but for other values of these parameters percentage of substance in node 0 can decrease.

Figure 3c shows influence of increasing value of the parameter $\hat{\gamma}_1$ on percentages of substance in the three nodes of channel. As there are no branches in studied channel then $\hat{\gamma}_1 = \gamma_1$. Parameter γ_1 accounts for the leakage from node 1 of channel. The increase of this leakage leads to decrease of percentage of substance in node 1 and in following node 2 at expense of percentage of substance in node 0 (node 0 is not affected by the leakage of substance in node 1 which position is after node 0).

Figure 3d shows influence of increasing value of the parameter $\hat{\gamma}_2$ on percentage of substance in nodes 0, 1, and 2. Parameter $\hat{\gamma}_2$ accounts for leakage of substance from node 2 of studied channel. There are no branches in studied channel and because of this $\hat{\gamma}_2 = \gamma_2$. Increased value of $\hat{\gamma}_2$ (increased leakage from node 2) leads to decrease of percentage of substance in node 2 and to corresponding increase of substance in nodes 0 and 1.

Figure 4 shows influence of increasing value of the parameter β_1 on probabilities $p_{0,1,2}$. Parameter β_1 accounts for additional outflow of substance from node 1 to node 2 because of some extra reason (in the theory of migration this extra reason can be preference of migrants which prefer to migrate to country 2 instead to stay in country 1). Increased value of β_1 leads to decrease of percentage of substance in the node 1 and to increase of percentage of substance which is located in node 2. For the values of parameters as in Figure 4 there is an additional change: percentage of substance in node 0 increases too.

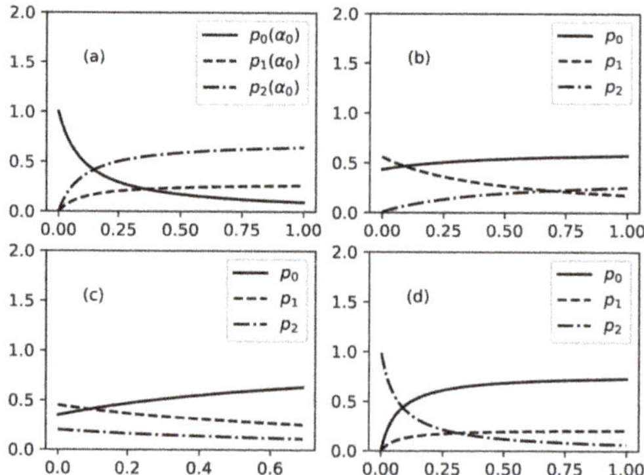

Figure 3. Probabilities p_0 (solid lines), p_1 (dashed lines), and p_2 (dot-dashed lines) as functions of selected parameter of the problem when all other parameters are fixed. (**a**): $p_0(\alpha_0)$, $p_1(\alpha_0)$; $p_2(\alpha_0)$ for fixed values of other parameters as follows: $\alpha_1 = 0.04$, $\beta_1 = 0.28$, $\hat{\gamma}_1 = 0.04$, $\hat{\gamma}_2 = 0.13$. (**b**): $p_0(\alpha_1)$, $p_1(\alpha_1)$; $p_2(\alpha_1)$ for fixed values of other parameters as follows: $\alpha_0 = 0.4$, $\beta_1 = 0.01$, $\hat{\gamma}_1 = 0.3$, $\hat{\gamma}_2 = 0.7$. (**c**): $p_0(\hat{\gamma}_1)$, $p_1(\hat{\gamma}_1)$; $p_2(\hat{\gamma}_1)$ for fixed values of other parameters as follows: $\alpha_0 = 0.4$, $\alpha_1 = 0.3$, $\beta_1 = 0.01$, $\hat{\gamma}_2 = 0.7$. (**d**): $p_0(\hat{\gamma}_2)$, $p_1(\hat{\gamma}_2)$; $p_2(\hat{\gamma}_2)$ for fixed values of other parameters as follows: $\alpha_0 = 0.2$, $\alpha_1 = 0.3$, $\beta_1 = 0.01$, $\hat{\gamma}_1 = 0.4$.

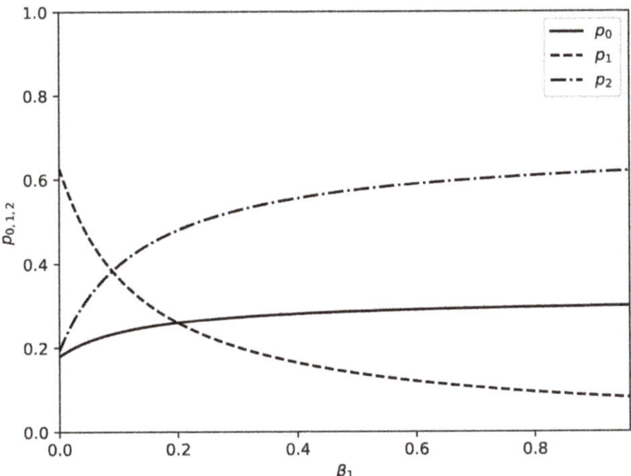

Figure 4. Probabilities $p_0(\beta_1)$ (solid lines), $p_1(\beta_1)$ (dashed lines), and $p_2(\beta_1)$ (dot-dashed lines) as functions of β_1 when all other parameters are fixed as follows: $\alpha_0 = 0.28$, $\alpha_1 = 0.04$, $\hat{\gamma}_1 = 0.04$, $\hat{\gamma}_2 = 0.13$.

Figures 5 and 6 show influence of changing values of parameters of problem on the values of information measures I_{p_0}, I_{p_1}, and I_{p_2} for nodes of studied channel. Figure 5a shows changes in the information measures with increasing value of parameter α_0 when values of all other parameters of problem are fixed. We observe that the value of I_{p_0} increases with increasing value of α_0 and values of I_{p_1} and I_{p_2} decrease with increasing value of α_1. This is because of the redistribution of percentage of substance in nodes 0, 1 and 2 with increasing value of α_0. Because of increasing outflow from node 0 the percentage of total substance located in this node decreases. The event associated with information measure I for node 0 becomes rare and occurrence of this event carries larger information. The percentage of substance in nodes 1 and 2 increases with increasing value of α_0. The event associated with information measure I becomes more frequent and this leads to decreasing information associated with occurrence of this event in nodes 1 and 2. Similar is the situation in Figure 5b where increasing value of parameter α_1 (accounting for outflow of substance from node 1 to node 2) leads to increasing percentage of substance in nodes 0 and 2 and decreasing percentage of substance in node 2. The information associated with occurrence of event of interest in node 2 increases and information associated with occurrence of event of interest in nodes 0 and 2 decreases.

Figure 5c shows influence of increasing value of the leakage parameter $\hat{\gamma}_1*$ (accounting for leakage from node 1) on information associated with occurrence of the event of interest in nodes 0, 1, and 2. Decreasing percentage of amount of substance in nodes 1 and 2 and increasing percentage of substance in node 0 lead to increasing information associated with occurrence of the event in nodes 1 and 2 and decreasing information associated with occurrence of event of interest in node 0. Situation connected to increasing value of the leakage parameter $\hat{\gamma}_2$ is shown in Figure 5d. This situation is similar to the situation from Figure 5c: increasing percentage of substance leads to decreasing value of information measure associated with occurrence of event of interest and decreasing percentage of substance leads to increasing value of information measure associated with occurrence of event of interest.

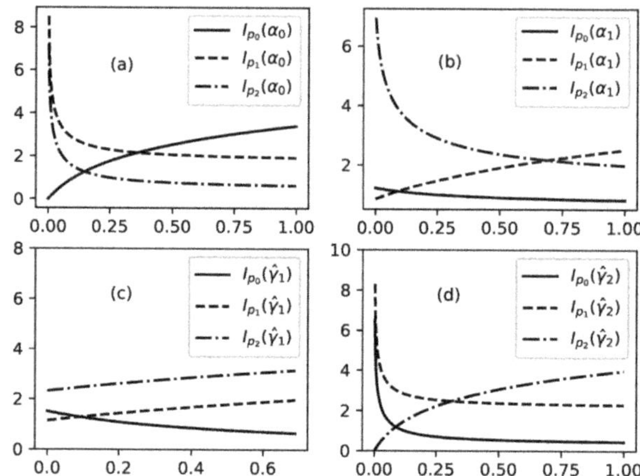

Figure 5. Information measures I_{p_0} (solid lines), I_{p_1} (dashed lines), and I_{p_2} (dot-dashed lines) as functions of selected parameter of the problem when all other parameters are fixed. (**a**): Dependence of I_{p_0}, I_{p_1}; I_{p_2} on α_0 for fixed values of other parameters as follows: $\alpha_1 = 0.04$, $\beta_1 = 0.28$, $\hat{\gamma}_1 = 0.04$, $\hat{\gamma}_2 = 0.13$. (**b**): Dependence of I_{p_0}, I_{p_1}; I_{p_2} on α_1 for fixed values of other parameters as follows: $\alpha_0 = 0.4$, $\beta_1 = 0.01$, $\hat{\gamma}_1 = 0.3$, $\hat{\gamma}_2 = 0.7$. (**c**): Dependence of I_{p_0}, I_{p_1}; I_{p_2} on $\hat{\gamma}_1$ for fixed values of other parameters as follows: $\alpha_0 = 0.4$, $\alpha_1 = 0.3$, $\beta_1 = 0.01$, $\hat{\gamma}_2 = 0.7$. (**d**): Dependence of I_{p_0}, I_{p_1}; I_{p_2} on $\hat{\gamma}_2$ for fixed values of other parameters as follows: $\alpha_0 = 0.2$, $\alpha_1 = 0.3$, $\beta_1 = 0.01$, $\hat{\gamma}_1 = 0.4$.

Figure 6 shows influence of increasing value of preference parameter β_1 on information measures I_{p_0}, I_{p_1}, and I_{p_2}. For corresponding fixed values of other parameters the increase of value of β_1 leads to increase of percentage of substance in nodes 0 and 2 and decrease of percentage of total substance located in node 2. The information associated with event of interest increases for events occurrence in node 2 and decreases for events occurrence in other two nodes.

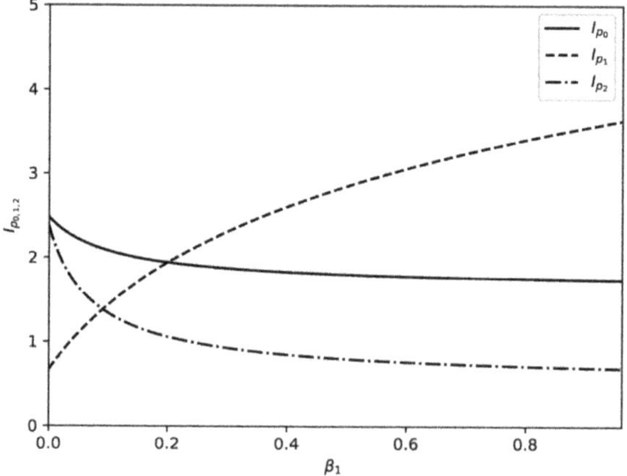

Figure 6. I_{p_0} (solid lines), I_{p_1} (dashed lines), and I_{p_2} (dot-dashed lines) as functions of β_1 when all other parameters are fixed as follows: $\alpha_0 = 0.28$, $\alpha_1 = 0.04$, $\hat{\gamma}_1 = 0.04$, $\hat{\gamma}_2 = 0.13$.

Finally Figure 7 shows influence of increasing value of selected parameters on the Shannon information measure for entire channel. Shannon information measure H is the average information we get from a occurrence of event of interest in nodes of studied channel. Two kinds of behavior of Shannon information measure are shown in Figure 7. First of all increase of value of selected parameter (with fixed values of other parameters) can lead to a maximum of the value of Shannon information measure for some value of changing parameter as shown in Figure 7a,b,d. Second kind of behavior is shown in Figure 7c where increasing value of leakage parameter γ_1 leads to monotonous decrease of value of the Shannon information measure.

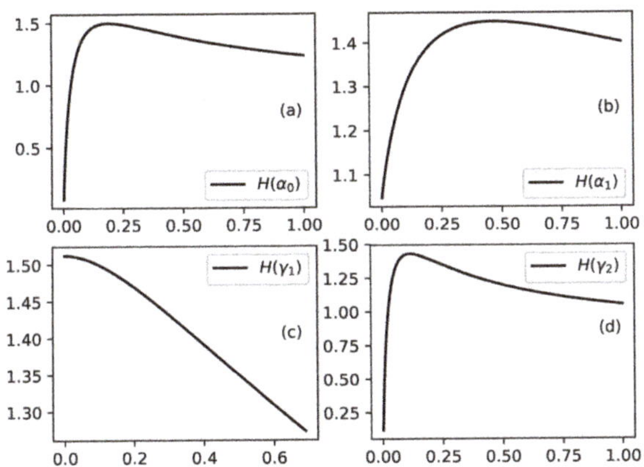

Figure 7. Shannon information measure as function of selected parameters when the other parameters of probability distribution are fixed. (**a**): $H(\alpha_0)$ for fixed values of other parameters as follows: $\alpha_1 = 0.04$, $\beta_1 = 0.28$, $\hat{\gamma}_1 = 0.04$, $\hat{\gamma}_2 = 0.13$. (**b**): $H(\alpha_1)$ for fixed values of other parameters as follows: $\alpha_0 = 0.4$, $\beta_1 = 0.01$, $\hat{\gamma}_1 = 0.3$, $\hat{\gamma}_2 = 0.7$. (**c**): $H(\hat{\gamma}_1)$ for fixed values of other parameters as follows: $\alpha_0 = 0.4$, $\alpha_1 = 0.3$, $\beta_1 = 0.01$, $\hat{\gamma}_2 = 0.7$. (**d**): $H(\hat{\gamma}_2)$ for fixed values of other parameters as follows: $\alpha_0 = 0.2$, $\alpha_1 = 0.3$, $\beta_1 = 0.01$, $\hat{\gamma}_1 = 0.4$.

4. Discussion

The results obtained above allow us to discuss various kinds of probability distributions. The conventional probability distributions correspond to a channel which has a single arm. Such distributions have been discussed in our previous work [23,26–28]. These distributions can be connected to Waring distribution, Zipf distribution, Yule-Simon distribution, Binomial distribution, etc. We can study also other kinds of distributions. One example is the probability distribution connected to distribution of substance in a channel which has more than one arm. More complicated case is the probability distribution connected to distribution of substance in a part of the studied network that contains several channels for motion of substance. The most complicated kind of probability distribution is the probability distribution connected to distribution of substance in all nodes of studied network.

5. Concluding Remarks

We discuss above a model of directed motion of substance through a channel of a network. The study is devoted to the stationary regime of motion of substance through channel arms and main outcomes are obtained new distributions connected to distribution of substance in nodes of channel. The model is formulated in such a way that it can have broad range of applicability. For an example the model can be used for study of motion of substance in technological systems or for study of motion of resources in various networks (e.g., motion of goods in logistic networks). The model can be applied also for study of other systems such as channels of human migration. Let us finish the text by an interpretation of obtained results from point of view of migration flows. For the case of migration flow migrants move through countries which form a migration channel and some migrants obtain permission to stay in corresponding country (which corresponds to leakage phenomenon in our model). Figure 3 shows the influence of model parameters on distribution of migrants in corresponding channel. Especially interesting is the influence of increasing leakage parameter shown in Figure 3d. Increasing leakage γ_2 means an increase of number of migrants who obtain permission to stay in the third country of studied channel. This can lead to drop of percentage of migrants without permission to stay in this country at the expense of the percentage of migrants without permission to stay in the other two countries of channel. Figure 3c shows that increasing leakage in second country of studied channel affect the percentage of migrants in the third country of the channel. If migrants have preferences for the third country of studied channel then percentage of migrants in this country increases with increased preference mostly at expense of percentage of migrants in previous country of channel—Figure 4. Changes of parameters of the model affect information about events connected to flows of migrants (e.g., information about criminal events). With increasing permeability of borders between countries (for an example with increasing value of parameter α_0) the amount of information connected to studied class of events increases in first country of studied channel and decreases in next two countries—Figure 5a. Opposite effect connected to increasing of leakage is shown in Figure 5c. Increasing value of preference parameter β_1 leads to increasing value of information connected with studied class of events in second country of the channel—Figure 6. Interesting is that the Shannon information measure connected to studied class of events can have maxima for selected values of model parameters—Figure 7.

Author Contributions: Conceptualization, N.K.V. and R.B.; methodology, Z.I.D. and N.K.V.; software, R.B.; validation, N.K.V., R.B. and Z.I.D.; formal analysis, R.B., Z.I.D. and N.K.V.; writing—original draft preparation, N.K.V.; writing—review and editing, N.K.V.; visualization, Z.I.D. and R.B.; supervision, N.K.V.; funding acquisition, N.K.V. All authors have read and agreed to the published version of the manuscript.

Funding: This research was partially supported by the project BG05 M2OP001-1.001-0008 "National Center for Mechatronics and Clean Technologies", funded by the Operating Program "Science and Education for Intelligent Growth" of Republic of Bulgaria and by the National Scientific Program "Information and Communication Technologies for a Single Digital Market in Science, Education and Security" (ICTinSES), contract No D01205/23.11.2018, financed by the Ministry of Education and Science in Bulgaria.

Conflicts of Interest: The authors declare no conflict of interest.

References

1. Marsan, G.A.; Bellomo, N.; Tosin, A. *Complex Systems and Society: Modeling and Simulation*; Springer: New York, NY, USA, 2013; ISBN 978-1-4614-7241-4.
2. Amaral, L.A.N.; Ottino, J.M. Complex Networks. Augmenting and Framework for the Study of Complex Systems. *Eur. Phys. J. B* **2004**, *38*, 147–162. [CrossRef]
3. Vitanov, N.K. *Science Dynamics and Research Production. Indicators, Indexes, Statistical Laws and Mathematical Models*; Springer: Cham, Switzerland, 2016; ISBN 978-3-319-41629-8.
4. Blasius, B.; Kurts, J.; Stone, L. (Eds.) *Complex Population Dynamics. Nonlinear Modeling in Ecology, Epidemiology and Genetics*; World Scientific: Singapore, 2007; ISBN 978-9-812-77157-5.
5. Vitanov, N.K.; Dimitrova, Z.I.; Ausloos, M. Verhulst-Lotka-Volterra Model of Ideological Struggle. *Physica A* **2010**, *389*, 4970–4980. [CrossRef]
6. Boccaletti, S.; Latora, V.; Moreno, Y.; Chavez, M.; Hwang, D.U. Complex Networks: Structure and Dynamics. *Phys. Rep.* **2006**, *424*, 175–308. [CrossRef]
7. Bertin, F. *Statistical Physics of Complex Systems*; Springer: Charm, Switzerland, 2016; ISBN 978-3-319-42338-8.
8. Vitanov, N.K.; Dimitrova, Z.I.; Vitanov, K.N. Traveling Waves and Statistical Distributions Connected to Systems of Interacting Populations. *Comput. Math. Appl.* **2013**, *66*, 1666–1684. [CrossRef]
9. Pastor-Sattoras, R.; Vespignani, A. Epidemic Dynamics and Endemic States in Complex Networks. *Phys. Rev. E* **2001**, *63*, 066117. [CrossRef]

10. Vitanov, N.K.; Ausloos, M.; Rotundo, G. Discrete Model of Ideological Struggle Accounting for Migration. *Adv. Complex Syst.* **2012**, *15*, 1250049. [CrossRef]

11. Ricard, J. *Biological Complexity and the Dynamics of Life Processes*; Elsevier: Amsterdam, The Netherlands, 1999; ISBN 978-0-444-50081-6.

12. Kalyagin, V.A.; Pardalos, P.M.; Rassias, T.M. (Eds.) *Network Models in Economics and Finance*; Springer: Charm, Switzerland, 2014; ISBN 978-3-319-09682-7.

13. Nakagawa, S.; Shikano, K.; Tohkura, Y. *Speech, Hearing and Neural Network Models*; IOS Press: Amsterdam, The Netherlands, 1995; ISBN 978-90-5199-178-9.

14. Castillo, E.; Gutierrez, J.M.; Hadi, A.S. *Expert Systems and Probabilistic Network Models*; Springer: New York, NY, USA, 1997; ISBN 978-1-4612-7481-0.

15. Ramos, P.P. *Network Models for Organizations*; Palgrawe Makmillan: New York, NY, USA, 2012; ISBN 978-0-230-32016-1.

16. Carrington, P.J.; Scott, J.; Wasserman, S. *Models and Methods in Social Network Analysis*; Cambridge University Press: Cambridge, UK, 2005; ISBN 978-0-511-81139-5.

17. Chan, W.-K. *Theory of Nets: Flows in Networks*; Wiley: New York, NY, USA, 1990; ISBN 978-0-471-85148-6.

18. Albert, R.; Barabasi, A.-L. Statistical Mechanics of Complex Networks. *Rev. Mod. Phys.* **2002**, *74*, 47–97. [CrossRef]

19. Dorogovtsev, S.N.; Mendes, J.F.F. Evolution of networks. *Adv. Phys.* **2002**, *51*, 1079–1187. [CrossRef]

20. Ford, L.D., Jr.; Fulkerson, D.R. *Flows in Networks*; Princeton University Press: Princeton, NJ, USA, 1962; ISBN 978-0-691-14667-6.

21. Harris, J.R.; Todaro, M.O. Migration, Unemployment and Development: A Two-Sector Analysis. *Am. Econ. Rev.* **1970**, *60*, 126–142.

22. Fawcet, J.T. Networks, Linkages, and Migration Systems. *Int. Migr. Rev.* **1989**, *23*, 671–680. [CrossRef]

23. Vitanov, N.K.; Vitanov, K.N. Box Model of Migration Channels. *Math. Soc. Sci.* **2016**, *80*, 108–114. [CrossRef]

24. Schubert, A.; Glänzel, W. A Dynamic Look at a Class of Skew Distributions. A model with Scientometric Applications. *Scientometrics* **1984**, *6*, 149–167. [CrossRef]

25. Gartner, N.H.; Imrota, G. (Eds.) *Urban Traffic Networks. Dynamic Flow Modeling and Control*; Springer: Berlin, Germany, 1995; ISBN 978-3-642-79643-2.

26. Vitanov, N.K.; Vitanov, K.N. Discrete-Time Model for a Motion of Substance in a Channel of a Network with Application to Channels of Human Migration. *Physica A* **2018**, *509*, 635–650. [CrossRef]

27. Vitanov, N.K.; Vitanov, K.N. On the Motion of Substance in a Channel of a Network and Human Migration. *Physica A* **2018**, *490*, 1277–1294. [CrossRef]

28. Vitanov, N.K.; Vitanov, K.N. Statistical Distributions Connected to Motion of Substance in a Channel of a Network. *Physica A* **2019**, *527*, 121174. [CrossRef]

29. Vitanov, N.K.; Vitanov, K.N.; Ivanova, T. Box Model of Migration in Channels of Migration Networks. *Adv. Comput. Ind. Math.* **2018**, *728*, 203–215. [CrossRef]

30. Vitanov, N.K.; Borisov, R. A Model of a Motion of Substance in a Channel of a Network. *J. Theor. Appl. Mech.* **2018**, *48*, 74–84. [CrossRef]

31. Borisov, R.; Vitanov, N.K. Human Migration: Model of a Migration Channel with a Secondary and a Tertiary Arm. *AIP Conf. Proc.* **2019**, *2075*, 150001. [CrossRef]

Article

Cross-Entropy as a Metric for the Robustness of Drone Swarms

Piotr Cofta, Damian Ledziński, Sandra Śmigiel and Marta Gackowska *

Faculty of Telecommunications, Computer Science and Technology, UTP University of Science and Technology, 85-796 Bydgoszcz, Poland; piotr.cofta@utp.edu.pl (P.C.); damian.ledzinski@utp.edu.pl (D.L.); sandra.smigiel@utp.edu.pl (S.Ś.)
* Correspondence: margac001@utp.edu.pl

Received: 21 April 2020; Accepted: 25 May 2020; Published: 27 May 2020

Abstract: Due to their growing number and increasing autonomy, drones and drone swarms are equipped with sophisticated algorithms that help them achieve mission objectives. Such algorithms vary in their quality such that their comparison requires a metric that would allow for their correct assessment. The novelty of this paper lies in analysing, defining and applying the construct of cross-entropy, known from thermodynamics and information theory, to swarms. It can be used as a synthetic measure of the robustness of algorithms that can control swarms in the case of obstacles and unforeseen problems. Based on this, robustness may be an important aspect of the overall quality. This paper presents the necessary formalisation and applies it to a few examples, based on generalised unexpected behaviour and the results of collision avoidance algorithms used to react to obstacles.

Keywords: entropy; cross-entropy; drones; swarms; robustness

1. Introduction

The development of drones and their swarms will eventually lead to crowded skies, particularly in urban environments. The safety of such environments depends on the way the behaviour of swarms is organised, including the need to keep them within allocated airspace. Furthermore, an appropriate organisation of swarms may have a positive impact on the fulfilment of their missions while requiring limited use of resources and minimising the impact on the environment.

However, the physicality of flight means that the swarm is always affected by some level of disorganisation, whether it is caused by changeable weather or unanticipated objects crossing the flight path. A certain level of disorganisation can be managed but excessive disorganisation may lead to significant damage. It is possible to calculate an acceptable disorganisation profile as a part of mission risk management.

This paper proposes the use of cross-entropy as a metric of the robustness of the swarm control algorithm, where the swarm is treated as a Shannon stochastic information source that is optimised for the acceptable level of disorganisation. Knowing the divergence from the acceptable, referential entropy will help control missions to avoid unacceptable levels of disorganisation and to compare mission control algorithms to identify those that prevent excessive disorganisation. This paper presents research that provides a proposition and some cases to support it. Further work is planned to verify and apply this theory to actual swarms of drones.

This paper starts with a brief note on terminology and goes on to introduce the model of the swarm. Further, the necessary formalisation with brief examples and discussion is presented. Then, an extended example is presented, as well as an overview of some related works with conclusions.

2. Terminology

A drone is a general term related to unmanned vehicles of various kinds [1], whether they be remotely operated or autonomous. For this paper, it is beneficial to primarily think of drones in terms of popular multirotor unmanned aerial vehicles, such as quadcopters. A swarm is a collection of drones under a single management system, occupying a certain space, interacting with each other, and pursuing their collective objective while avoiding collisions (we intentionally exclude swarms whose intention is to engage in a collision). A mission is any time during which drones in a swarm move to a specific target while performing the desired trajectory. A mission should be realised optimally and safely. A mission can be performed automatically, semi-automatically or autonomously, depending on the management methods.

3. Model

Let us consider a swarm of drones $D = \{d_i; i = 1, \ldots, k\}$ that execute mission $M = \{t_j; j = 1, \ldots, m; t_{j+1} > t_j\}$, which is defined as an ordered, evenly spaced, sequence of moments in time. To accomplish the mission, drones must progress through a series of respective states, e.g., follow certain trajectories. As long as drones follow their planned trajectories, the swarm is considered to be organised. Unexpected events, such as changes in the weather or the intrusion of objects into flight paths, make the swarm diverge and introduce some degree of disorganisation to the otherwise organised structure of a swarm.

The swarm can withstand disorganisation up to a certain level. While such a level can be defined in different ways, here it is described by an overall disorganisation "mass", where each divergence contributes to such a mass. Events of low-impact divergences (e.g., being slightly off course) have a small contribution, whereas events of high-impact divergences (e.g., leaving the allocated perimeter) have a large contribution. Certain substitutions are possible, e.g., a few low-impact events can be considered to be the equivalent of a single high-impact one.

Based on the relative occurrence of different events, it is possible to construct the acceptable probability of events such that low-impact events can appear more frequently than high-impact ones. Note that some impact is unavoidable, there is no state free from some disorganisation. Considering that there is a finite precision, the state of being perfectly on course can be determined even if such a state is not free from some impact.

As drones report their behaviour through events, it is possible to determine the difference between the acceptable probability distribution of events and the actual one to find out whether the mission flies at low or high levels of disorganisation.

4. Formalisation

Let us assume that there is a set of n classes of possible states, each reflecting a certain level of divergence such that at any time interval, the drone can report one of n possible events. Those classes (and associated events) are identified as $C = \{c_i, i = 1, \ldots, n\}$. At regular intervals throughout the mission, each drone communicates the class they are currently in. With every class, there is an associated relative impact factor on disorganisation, $F = \{f_i, i = 1, \ldots, n\}$. The assessment of such an impact can be provided, e.g., by the analysis of previous missions in a way similar to risk assessment.

From the relationship between various impact factors, it is possible to calculate the normalised discrete probability distribution Q, where events from classes associated with the lower impact factor are granted a higher probability of occurrence. For example:

$$Q = \{q_i \; ; \; i = 1, \ldots, n\}; q_i = \frac{1}{f_i \times \sum_{j=1,\ldots,n} \frac{1}{f_j}}. \tag{1}$$

It is also possible to construct Q using other methods, e.g., by sampling previous events in a manner that is often used in risk management. As high-impact events also tend to be low probability events, efficient sampling may be based on importance and may even internally employ cross-entropy to determine Q [2].

The distribution Q describes the referential probability distribution when the overall disorganisation is still at an acceptable level. This distribution is associated with a referential entropy, i.e., a level of disorganisation that does not disrupt the mission. This referential entropy can be calculated using Shannon's equation [3]:

$$H(q) = -\sum_{i=1,\ldots,\,n} q_i \times \log(q_i).$$

(2)

As drones communicate signals, it is possible to determine the probability distribution of receiving various classes of signals, $P = \{p_i,\ i = 1, \ldots, n\}$. Based on this, the cross-entropy between the observed and the referential entropy can be calculated using [4]:

$$H(p, q) = -\sum_{i=1,\ldots,\,n} p(x_i) \times \log(q(x_i)).$$

(3)

This cross-entropy can be interpreted as a measure of disorganisation of the swarm relative to the referential one. Note that the cross-entropy can be both smaller and larger than the referential one, indicating situations of (acceptable) low disorganisation and of high (potentially unacceptable) disorganisation, respectively.

For clarification, the levels of entropy that are above the referential one represent an increased risk to the mission but do not necessarily signify its failure. In practice, the levels below the referential one may allow the swarm to continue its operation in a fully automatic or even autonomous manner, while levels higher than the referential one may call for an operator's action.

For example, let us consider a swarm that can emit five classes of signals that represent five real-life situations, namely c_1: the drone is on course, c_2: the drone is slightly off course but not disturbing other drones, c_3: the drone is within the limits of the swarm but it is disturbing other drones, c_4: the drone has left the perimeter of the swarm and c_5: the drone has lost contact with the swarm.

The impact of various classes on the disorganisation has been determined and shown in Table 1. Such an impact can be associated, e.g., with the mission average delays, energy usage, or other risk factors. Note that it is only the relative impact that is important, not the absolute values.

Table 1. Relative impact of classes of events.

c_1	c_2	c_3	c_4	c_5
0.01	1.0	20.0	50.0	100.0

This implies the following (Table 2) probability distribution (all values rounded).

Table 2. Probability distribution for classes of events.

q_1	q_2	q_3	q_4	q_5
0.9893	0.0990	0.0005	0.0002	0.0001

The referential entropy calculated according to the equation above is approx. 0.090. Therefore, if the actual entropy of the swarm is below this value, the swarm can be characterised as having low disorganisation, while higher values indicate an extent of disorganisation that may unacceptably increase the risk to the mission.

Case 1: Let us consider a situation where the mission has been disturbed such that several drones left their planned trajectories. The observed distribution of events is as shown in Table 3:

Table 3. Probability distribution for significant disturbances.

p_1	p_2	p_3	p_4	p_5
0.8	0.1	0.06	0.03	0.01

For this case, the cross-entropy is approx. 1.840, much higher than the referential one. This indicates an increase in disorganisation and an increased risk.

Case 2: Let us consider the same swarm flying a very quiet mission where the drones fly exactly as planned all the time. The observed distribution P has the form of the Kronecker delta (Table 4).

Table 4. Probability distribution for the undisturbed mission.

p_1	p_2	p_3	p_4	p_5
1.0	0.0	0.0	0.0	0.0

In this case, the cross-entropy is approx. 0.016, lower than the referential one. This indicates that this mission is well organised.

5. Continuous and Mixed Probability Distributions

As the size and the density of swarms grow, it is reasonable to consider a continuous case where both probability distributions P and Q are continuous functions over some support X. For example, probability distribution Q may be derived from a continuous function linking the distance from the correct state with the impact on disorganisation. Alternatively, distribution P can be a continuous estimate of the actual discrete distribution.

$$H(p,q) = -\int_{x \in X} P(x) \times \log Q(x)\, dx \tag{4}$$

Situations of mixed distributions, where Q is continuous while p is discrete, are of some practical use. For example, the drone management system may define the continuous impact function that links the distance between the expected and the actual state to the extent of the impact, resulting in a continuous distribution of Q. Meanwhile p can be discrete, as it is being calculated from events observed during the swarm mission.

For such situations, the cross-entropy can be determined as follows, where $Q(x)$ is the value of the probability distribution function calculated using the impact function for given value x:

$$H(p,q) = -\sum_{x \in X} p(x) \times \log(Q(x)). \tag{5}$$

An alternative approach may require the use of the fixed-width quantisation of functions over the support in the form of an LDDP (limiting density of discrete points). This may introduce the need for correction of the quantisation error [5,6]. Note that the use of cross-entropy in this paper is a comparative one: it is more interesting to compare two values than present their correctness. Consequently, both approaches may provide a solution.

6. Discussion

6.1. Advantages of Cross-Entropy

The difference between the desired and the actual state of drones can be expressed through various loss metrics, specifically using the mean squared error (MSE) or through cross-entropy. The choice of cross-entropy over MSE comes from the intended purpose of the loss metric, namely to improve the control algorithm for the swarm of drones. Compared to MSE, cross-entropy stresses the small differences from the referential level of entropy, i.e., cross-entropy allows for better differentiation between what is acceptable (in terms of disorganisation) and what is excessive.

The problem of drones diverging from their intended trajectories is mostly seen as a control problem with an impact on energy loss. Whatever method is used to manage the swarm, it has to determine its situation and undertake control measures to revert drones to their routes. Thus, the key determinants of the extent of the divergence are the control precision and energy loss. Out of these two loss metrics, cross-entropy that is a better estimator for the required additional amount of information.

From the perspective of control, small differences must not be neglected, as they still require actions to be taken and the swarm control algorithm must be adjusted to react to them without over-reacting. In contrast, when the difference is large, the extent of this difference gradually matters less from the perspective of control, as only some coarse actions have to be taken.

Similarly, from the perspective of energy consumption and owing to the inertia of the drone, small manoeuvres are relatively expensive (per the unit of distance), while large manoeuvres can be inexpensive. Thus, the cross-entropy seems to be a better estimator for both the control overhead and the use of energy.

6.2. Coding Scheme

Cross-entropy is susceptible to the choice of the coding scheme (i.e., the distribution q). Specifically, q must anticipate all possible classes of signals by assigning them certain non-zero probabilities over all the support of p. If such a distribution is derived from the impact factor, then no signal is allowed to be free from the impact. Otherwise, if the swarm is in situations not anticipated by q (i.e., where the distribution is either undefined or zero), the information content of such signals is undefined or infinite.

The construction of an appropriate q can be done using different methods; this paper does not assume that the impact factor is the only appropriate one. The impact factor is conceptually similar to methods used in machine learning, where there are weighted penalties for misclassification [7]. The authors' choice of a method for the determination of the probability distribution was affected by two factors. First, the probability distribution forms the common denominator for various methods to determine q. Second, the problem of controlling the swarm is a signalling and information problem; therefore, adherence to the cross-entropy origin was preferential.

There is a further similarity between the problem of constructing q and machine learning where a training set does not have samples of some classes, leading to difficulties in calculating the cross-entropy. This is improved by the recommendation to include samples for all classes. Similarly, this paper contains a restriction that there is no class without a certain impact. Note that, formally, this approach is justified: in real life, the class of "being perfectly on course" cannot be described as a Dirac delta due to measurement imprecision. Hence, it must always contain some range of low-impact behaviours.

The problem of assigning an appropriate probability to events that are only anticipated or that are rare is known, with some propositions applying cross-entropy to optimise the importance of sampling [2]. This proposition does not improve on them, as it only assumes that q can be determined. The proposed use of a real-life impact function closely links the proposed metric with actual costs and benefits of various behaviours, without claiming any particular shape of the impact function.

6.3. Referential Entropy

This model uses a referential probability distribution and its referential entropy to calculate the cross-entropy. It makes cross-entropy both positive and negative, depending on whether the actual distribution of signals indicates that the level of disorganisation is higher or lower than the referential one. Consequently, in a way that is different than in other applications of cross-entropy, there is no minimum entropy level that the cross-entropy will always be higher than, even though such cross-entropy can be used to judge (and to optimise) the swarm management algorithm.

There is an underlying assumption that the swarm can take certain levels of disorganisation such that not every localised disorganisation has an immediate impact on the swarm. Specifically, that imprecision in measurements may not allow us to confirm whether any drone is actually in the desired state. Therefore, the algorithm quality should not be judged by absolute perfection but rather by staying on the safe side of the referential level of disorganisation.

This reasoning means that the Kullback–Leibler (K-L) divergence is not applicable to this case. K-L divergence provides an estimate of the difference between the entropy of the actual distribution and the cross-entropy. In this case, it is at its minimum when $p = q$, i.e., when the swarm is disorganised exactly in an acceptable way. Whether the actual distribution indicates more or less disorder than the swarm can bear, the K-L divergence will grow. Still, K-L divergence may be a useful metric of the distance from the referential disorganisation that helps design control algorithms.

7. Extended Example

Collision avoidance is an example of a process that temporarily increases the disorganisation of the swarm. That is, in the presence of a disturbance, drones have to veer off course and break the formation. In such situations, they find themselves at locations that are less expected. Intuitively, the entropy of a swarm should increase. However, such an increase should only be temporary. In the absence of continuous disturbances, drones should return to the desired paths to continue their mission. As the drones again appear at the expected locations, the entropy of the swarm should decrease. Thus, the managed swarm should be perceived as an entropy-minimisation device, i.e., its entropy will increase only to the extent required to overcome an obstacle and will decrease once the obstacle is removed. The quality of such a strategy is reflected by the level of entropy carried by the swarm throughout the mission.

Collision avoidance by unmanned aerial vehicles can be implemented in many ways. Gong et al. [8] used a gradient-based collision avoidance algorithm for the control of multi-agent formations. The algorithm uses a consensus theory and a graph theory applied to three topologies. This method uses two circular zones that define distances from adjacent objects, whether they be obstacles or neighbours, to define prohibited zones.

This section makes use of the collision avoidance algorithm developed by Cofta et al. [9] that bears a superficial similarity to the one described above. The algorithm itself is inspired by the physics of repulsive forces (see, e.g. [10]), thus algorithmically replicating the exclusivity. The algorithm is run by each drone independently for at least at fixed time intervals and always planning for the interval ahead. It considers the location and movement of neighbouring drones. If some drones are nearby, then the algorithm alters the paths of drones to avoid close encounters; otherwise, they attempt to continue their original mission. The equations that define the algorithm are as follows:

$$\vec{p}_z = \vec{p}_m + \tau * \vec{p}_u, \tag{6}$$

$$\vec{p}_i = \overrightarrow{(x_i, x_j)}, \tag{7}$$

$$i \in N \leftrightarrow \left|\vec{p}_i\right| \leq R \text{ and } i \neq j, \tag{8}$$

$$\vec{u}_i = \frac{\vec{p}_i}{\left|\vec{p}_i\right|}, \tag{9}$$

$$\vec{p}_u = \sum_{i \in N} \vec{u}_i (R - \left|\vec{p}_i\right|)^Q, \tag{10}$$

where:

$\vec{p}_z$—vector that the drone will adopt for the next time interval.

$\vec{p}_m$—mission vector, i.e., the vector that the drone should have adopted to continue the mission; this vector is calculated to achieve the objective of the mission in the current situation if it were ignoring obstacles.

$\vec{p}_u$—escape vector that should be adopted to escape close encounters with neighbours disregarding the mission.

x_j—drone's location.

x_i —the location of the neighbour i.

$\vec{u}_i$—unit vector of $\vec{p}_i$.

N—the set of neighbours within radius R.

R—radius within which neighbours are of interest (in this simulation $R = 25$ m).

Q and τ—parameter constants (in this simulation $Q = 1$ and $\tau = 1$).

The following examples were developed using the simulation software created by the authors. The probability function used in those examples directly relates the distance between the actual and expected locations of the drone. It is defined as the cumulative distribution function of the lambda distribution. Consequently, the probability distribution Q is a lambda function $Q(x) = \lambda e^{-\lambda x}$; $\lambda = 1$, with a referential entropy of 1. As the p distribution is a discrete one, the mixed version of cross-entropy is used, where the sum is run over the product of the discrete probability distribution p and the point value of $Q(x)$.

This approach allows for an introduction of "momentary" entropy, where the distribution p takes on the form of the Kronecker delta. This allows for analysing changes in the entropy over time in response to obstacles for an individual drone or a swarm of them. While the "momentary" entropy is not intended to be used as a metric of robustness, it is a useful tool to illustrate and analyse the behaviour of drones and swarms.

Video recordings of simulations related to the extended example are included as Supplementary Materials.

7.1. Close Encounter

Figure 1 shows the trajectory of the flight of a swarm D that consists of three drones. D_2 and D_3 fly to the east, while D_1 flies to the west. Their trajectories were close enough such that the collision avoidance algorithm was triggered for all of them.

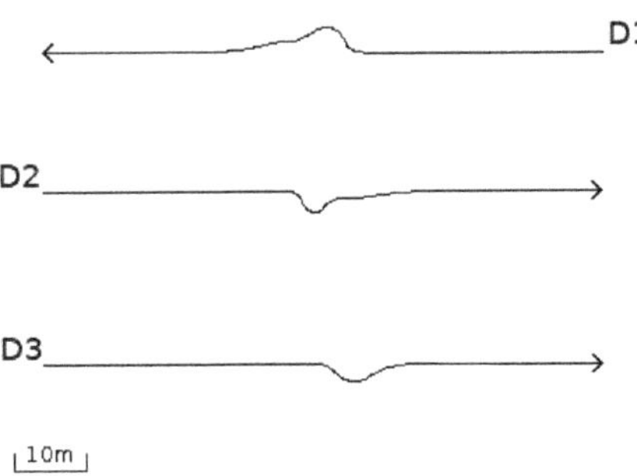

Figure 1. Close encounter of drones.

When the two topmost drones got too close to each other, they both veered off their courses, as expected. However, this caused the central drone to move closer to the lower one, triggering its algorithm as well. Eventually, all drones moved away from their courses. Once the distance became safe again, they returned to their original courses. Figure 2 shows the momentary entropy of each drone as a function of time.

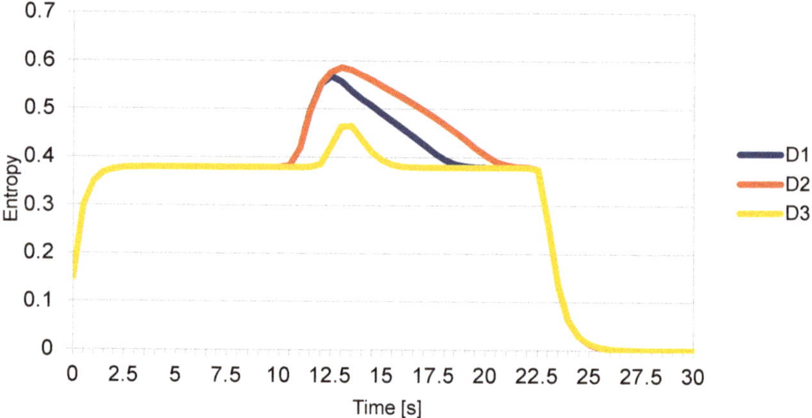

Figure 2. Momentary entropy of each drone as a function of time.

7.2. Grid Formation

Figure 3 shows a sixteen-drone swarm with drones stationary in a grid formation. The objective of their mission was to remain in this formation and at their current positions. The simulation depicted a situation where there was an intruder drone that passed through the swarm, paying no attention to other drones, to the extent of potentially colliding with them. However, each drone in the swarm was paying attention to every other drone, including the intruder, executing the collision avoidance algorithm. Note that the path of the intruder was straight, while drones from the swarm gave way. Once the intruder passed, each drone returned to its assumed position in the formation. Because the formation was relatively dense, the movement of one drone affected its neighbours.

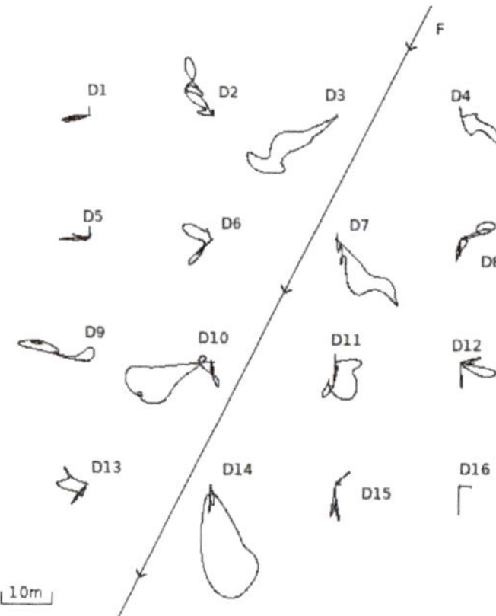

Figure 3. Sixteen-drone swarm with an initial grid formation. The swarm avoids the intruder (straight line) before returning to the grid formation via the paths shown.

Figure 4 shows the change in the momentary entropy over time of particular drones from the swarm. Figure 5 shows the total momentary entropy of the swarm over time. As expected, the appearance of the intruder initially increased the entropy but once the intruder left the swarm, it eventually returned to just above zero.

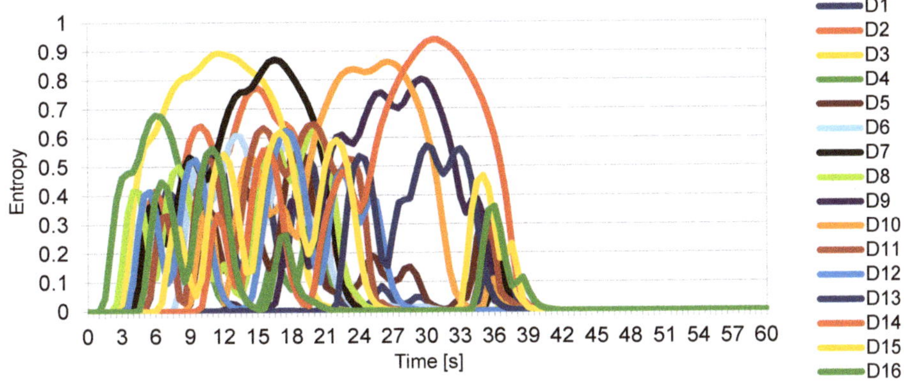

Figure 4. Momentary entropy of drones within the swarm.

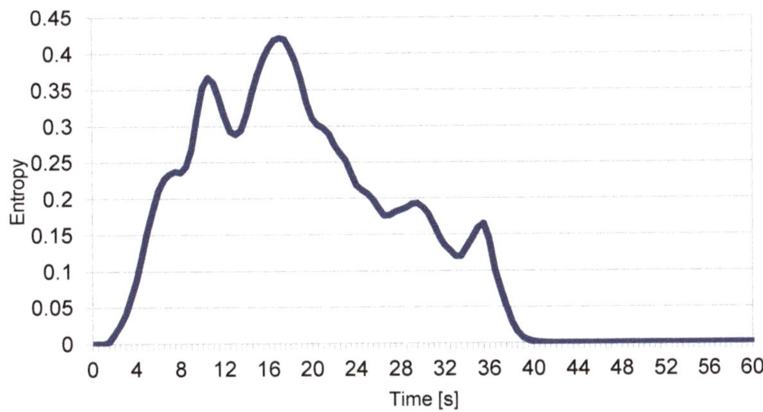

Figure 5. Momentary entropy of the swarm.

Note that the momentary entropy assumed the value of $-\log(Q(x))$, i.e., it could reach any positive value. The change in entropy over time, as shown in Figure 5, was calculated by including events from all drones for a given moment in time. As all drones reported events at regular intervals, this made the values presented in Figure 5 the average of the respective values from Figure 4.

Note that neither the momentary entropy of the drones nor the momentary entropy of the swarm ever exceeded the value of the referential entropy, which means that the level of disorganisation was acceptable. The entropy of the whole swarm, calculated for the whole passage of the intruder, was approx. 0.14. Again, this indicates that, despite the disturbances, the swarm was reasonably organised and that the mission itself was not endangered.

8. Related Works

Cross-entropy is a measure of a discrepancy between two probability distributions [4]. It is used widely beyond the theory of information, e.g., as an objective function for the optimisation of traffic flow [11] or in a particle swarm optimisation [12,13]. It is also used in machine learning as a loss function for the training set of neural networks [14] or to improve the clustering of data [15]. Further, it is used in robotics for the optimisation of controllers based on fuzzy logic [16,17]. Regarding the social sciences, it can also be used to explain complex global behaviours [18] and can be used in swarm intelligence [19].

Swarms and their self-organisation borrows much from the observation of bees [20], locusts [21], birds and fish [22–25]. The close resemblance between unmanned aircraft and insects or animals has been researched [26], specifically in terms of collision avoidance and in collaborative intelligence [27], while Can et al. [28] applies the rules of particle physics to swarms.

In those diverse areas, entropy is defined by re-applying Shannon's formula to various forms of grouping. For example, Folino and Forestiero [29], inspired by Van Dyke Parunak and Brueckner [30], demonstrates how entropy can be used to assess the properties of self-organising flocking algorithms by observing changes in entropy resulting from coupling organised and disorganised systems.

The application of cross-entropy to the process of training neural networks bears some resemblance to the problem discussed here, specifically for discrete distributions. Cross-entropy is one of the standard loss functions, specifically regarding multi-label classification [31]. Such a loss function can be extended to include some penalties for mislabelling [32,33], making it attractive for some real-world cases where misclassification should be penalised [6].

Particle swarm optimisation (PSO) [34] can use entropy for the simulated set of states ("particles") (EA-PSO) [12], and then it may apply cross-entropy in the meta-optimisation of the search space. Various modifications and extensions exist, such as memetic based [13], niche strategy [35], or clustering [36]. The evolutionary approach is used in Hu et al. [37], while Zhang et al. [38] employs direct competition.

PSO may prematurely converge to local optima since the best performing particle attracts the remaining ones. Therefore, the problem of diversity management (i.e., having particles exploring different alternatives beyond the local optimum) is important. Entropy is used in a way inspired by Shannon to manage the extent of diversity at the swarm level, combined with the optimum-seeking behaviour of particles at the local level.

Cross-entropy is used by PSO (among others) as a way to meta-optimise the solution space. For example, Yin [39] applies cross-entropy minimisation to determine the optimum threshold in image segmentation by comparing the probability distribution of the original image and the one after a threshold has been applied.

9. Conclusions

This paper proposes the use of cross-entropy as a metric for the quality of algorithms that manage swarms of drones. It reflects the extent of disorganisation of the swarm throughout its mission, where such disorganisation should be always minimised as much as possible. This cross-entropy is calculated relative to the referential probability distribution that is constructed out of the real-world impact that various events may have on the swarm.

Initial simulations demonstrated the viability of cross-entropy as a metric and allowed for distinguishing between missions with low and high levels of disorganisation. This is a work in progress. The authors plan to run both simulations and field experiments to determine the practical usefulness of the proposed metric under various flight conditions.

Supplementary Materials: The following are available online at http://www.mdpi.com/1099-4300/22/6/597/s1, Video S1: Grid, Video S2: Grid (debug), Video S3: Simple, Video S4: Simple (debug).

Author Contributions: Conceptualisation: P.C. and D.L.; methodology: P.C.; formal analysis: P.C.; supervision: D.L. and P.C.; software, validation and investigation: D.L.; writing—original draft preparation: P.C., M.G., S.Ś. and D.L.; writing—review and editing: S.Ś.; visualisation: D.L. and S.Ś.; project administration: M.G. All authors have read and agreed to the published version of the manuscript.

Funding: This research received no external funding.

Conflicts of Interest: The authors declare no conflict of interest.

References

1. Parlin, K.; Alam, M.M.; Le Moullec, Y. Jamming of UAV remote control systems using software defined radio. In Proceedings of the 2018 International Conference on Military Communications and Information Systems (ICMCIS), Warsaw, Poland, 22–23 May 2018.
2. Ridden, A.; Rubinstein, R. Minimum Cross-entropy Methods for Rare-event Simulation. *Simulation* **2007**, *83*, 769. [CrossRef]
3. Shannon, C.E. A mathematical theory of communication. *Bell Syst. Tech. J.* **1948**, *27*, 379–423. [CrossRef]
4. Rao, C.R. Entropy and Cross Entropy: Characterizations and Applications. In *The Legacy of Alladi Ramakrishnan in the Mathematical Sciences*; Alladi, K., Klauder, J., Rao, C., Eds.; Springer: New York, NY, USA, 2010. [CrossRef]
5. Jaynes, E.T. Information Theory and Statistical Mechanics. In *Statistical Physics (PDF)*; Ford, K., Ed.; Benjamin: New York, NY, USA, 1963; p. 181.
6. Jaynes, E.T. Prior Probabilities. *IEEE Trans. Syst. Sci. Cybern.* **1968**, *4*, 227–241. [CrossRef]
7. Mahdy, M. Weighted Entropy Measure: A New Measure of Information with its Properties in Reliability Theory and Stochastic Orders. *J. Stat. Theory Appl.* **2018**. [CrossRef]
8. Gong, Q.; Wang, C.; Qi, Z.; Ding, Z. Gradient-based collision avoidance algorithm for second-order multi-agent formation control. In Proceedings of the 2017 36th Chinese Control Conference (CCC), Dalian, China, 26–28 July 2017.
9. Cofta, P.; Ledziński, D.; Śmigiel, S.; Gackowska, M.; Jerks, N. Swarm Drone Communication & Collision Avoidance. Patent Application 431104, 10 September 2019.
10. Geiger, H.; Marsden, E. On a Diffuse Reflection of the α-Particles. *Proc. R. Soc. A* **1909**, *82*, 495–500. [CrossRef]
11. Ma, T.-Y.; Lebacque, J.-P. A Cross Entropy Based Multi-Agent Approach to Traffic Assignment Problems. *Traffic Granul. Flow* **2009**, *7*, 161–170. [CrossRef]
12. Guo, W.; Zhu, L.; Wang, L.; Wu, Q.; Kong, F. An Entropy-Assisted Particle Swarm Optimizer for Large-Scale Optimization Problem. *Mathematics* **2019**, *7*, 414. [CrossRef]
13. Petalas, Y.G.; Parsopoulos, K.E.; Vrahatis, M.N. Entropy–Based Memetic Particle Swarm Optimization for Computing Periodic Orbits of Nonlinear Mappings. In Proceedings of the 2007 IEEE Congress on Evolutionary Computation, Singapore, 25–28 September 2007.
14. Aurelio, Y.S.; de Almeida, G.M.; de Castro, C.L.; Braga, A.P. Learning from Imbalanced Data Sets with Weighted Cross-Entropy Function. *Neural Process. Lett.* **2019**, *50*, 1937–1949. [CrossRef]
15. Liu, B.; Pan, J.; McKay, R.I. Entropy-based metrics in swarm clustering. *Int. J. Intell. Syst.* **2009**, *24*, 989–1011. [CrossRef]
16. Anisimov, D.N.; Dang, T.S.; Banerjee, S.; Mai, T.A. Design and implementation of fuzzy-PD controller based on relation models: A cross-entropy optimization approach. *Eur. Phys. J. Spec. Top.* **2017**, *226*, 2393–2406. [CrossRef]
17. Olivares-Mendez, M.A.; Mejias, L.; Campoy, P.; Mellado-Bataller, I. Cross-Entropy Optimization for Scaling Factors of a Fuzzy Controller: A See-and-Avoid Approach for Unmanned Aerial Systems. *J. Intell. Robot. Syst.* **2012**, *69*, 189–205. [CrossRef]
18. Kordon, A.K. Swarm Intelligence: The Benefits of Swarms. In *Applying Computational Intelligence*; Springer: Berlin/Heidelberg, Germany, 2010. [CrossRef]
19. Blum, C.; Li, X. Swarm Intelligence in Optimization. In *Swarm Intelligence. Natural Computing Series*; Blum, C., Merkle, D., Eds.; Springer: Berlin/Heidelberg, Germany, 2008. [CrossRef]
20. Cully, S.M.; Seeley, T.D. Self-assemblage formation in a social insect: The protective curtain of a honey bee swarm. *Insectes Sociaux* **2004**, *51*, 317–324. [CrossRef]
21. Kurdi, H.A.; Aloboud, E.; Alalwan, M.; Alhassan, S.; Alotaibi, E.; Bautista, G.; How, J.P. Autonomous task allocation for multi-UAV systems based on the locust elastic behavior. *Appl. Soft Comput.* **2018**, *71*, 110–126. [CrossRef]
22. Parrish, J.K.; Viscido, S.V.; Grünbaum, D. Self-Organized Fish Schools: An Examination of Emergent Properties. *Biol. Bull.* **2002**, *202*, 296–305. [CrossRef] [PubMed]
23. Okubo, A. Dynamical aspects of animal grouping: Swarms, schools, flocks, and herds. *Adv. Biophys.* **1986**, *22*, 1–94. [CrossRef]

24. Toner, J.; Tu, Y. Flocks, herds, and schools: A quantitative theory of flocking. *Phys. Rev.* **1998**, *58*, 4828–4858. [CrossRef]

25. Reynolds, C.W. Flocks, herds and schools: A distributed behavioral model. In Proceedings of the 14th Annual Conference on Computer Graphics and Interactive Techniques (SIGGRAPH 1987), Association for Computing Machinery, New York, NY, USA, 27–31 July 1987; Volume 87, pp. 25–34. [CrossRef]

26. Janson, S.; Middendorf, M.; Beekman, M. Honeybee swarms: How do scouts guide a swarm of uninformed bees? *Anim. Behav.* **2005**, *70*, 349–358. [CrossRef]

27. Vásárhelyi, G.; Virágh, C.; Somorjai, G.; Nepusz, T.; Eiben, A.E.; Vicsek, T. Optimized flocking of autonomous drones in confined environments. *Sci. Robot.* **2018**, *3*, eaat3536. [CrossRef]

28. Can, C.F.; Bayram, Ç.; Toksoy, A.K.; Avsar, H.; Ozdemir, S. *Characterization of Swarm Beahavior through Pairwise Interactions by Tsallis Entropy*; IC-AI: Las Vegas, NV, USA, 2005; Volume 2.

29. Folino, G.; Forestiero, A. Using Entropy for Evaluating Swarm Intelligence Algorithms. In *Nature Inspired Cooperative Strategies for Optimization (NICSO 2010). Studies in Computational Intelligence*; González, J.R., Pelta, D.A., Cruz, C., Terrazas, G., Krasnogor, N., Eds.; Springer: Berlin/Heidelberg, Germany, 2010; Volume 284. [CrossRef]

30. Van Dyke Parunak, H.; Brueckner, S. Entropy and self-organization in multi-agent systems. In Proceedings of the Fifth International Conference on Autonomous Agents (AGENTS 2001), Montreal, QC, Canada, 28 May–1 June 2001; ACM: New York, NY, USA, 2001; pp. 124–130. [CrossRef]

31. Dong, Q.; Zhu, X.; Gong, S. Single-Label Multi-Class Image Classification by Deep Logistic Regression. *Proc. Conf. Artif. Intell.* **2019**, *33*, 3486–3493. [CrossRef]

32. Rengasamy, D.; Jafari, M.; Rothwell, B.; Chen, X.; Figueredo, G.P. Deep Learning with Dynamically Weighted Loss Function for Sensor-Based Prognostics and Health Management. *Sensors* **2020**, *20*, 723. [CrossRef]

33. Ho, Y.; Wookey, S. The Real-World-Weight Cross-Entropy Loss Function: Modeling the Costs of Mislabeling. *IEEE Access* **2020**, *8*, 4806–4813. [CrossRef]

34. Kennedy, J.; Eberhart, R.C. Particle swarm optimization. In Proceedings of the IEEE International Conference on Neural Networks, Perth, WA, Australia, 27 November–1 December 1995; IEEE: Perth, Australia, 1995; Volume 4, pp. 942–1948.

35. Li, D.; Guo, W.; Wang, L. Niching Particle Swarm Optimizer with Entropy-Based Exploration Strategy for Global Optimization. In *Advances in Swarm Intelligence. ICSI 2019. Lecture Notes in Computer Science*; Tan, Y., Shi, Y., Niu, B., Eds.; Springer: Cham, Switzerland, 2019; Volume 11655. [CrossRef]

36. Çomak, E. A modified particle swarm optimization algorithm using Renyi entropy-based clustering. *Neural Comput. Appl.* **2015**, *27*, 1381–1390. [CrossRef]

37. Hu, W.; Hu, J.; Zhang, X. The Improved Particle Swarm Optimization Based on Swarm Distribution Characteristics. In *Artificial Intelligence and Computational Intelligence. AICI 2012. Lecture Notes in Computer Science*; Lei, J., Wang, F.L., Deng, H., Miao, D., Eds.; Springer: Berlin/Heidelberg, Germany, 2012; Volume 7530. [CrossRef]

38. Zhang, W.-X.; Chen, W.-N.; Zhang, J. A dynamic competitive swarm optimizer based-on entropy for large scale optimization. In Proceedings of the 2016 Eighth International Conference on Advanced Computational Intelligence (ICACI), Chiang Mai, Thailand, 14–16 February 2016.

39. Yin, P.-Y. Multilevel minimum cross entropy threshold selection based on particle swarm optimization. *Appl. Math. Comput.* **2007**, *184*, 503–513. [CrossRef]

MDPI
St. Alban-Anlage 66
4052 Basel
Switzerland
Tel. +41 61 683 77 34
Fax +41 61 302 89 18
www.mdpi.com

Entropy Editorial Office
E-mail: entropy@mdpi.com
www.mdpi.com/journal/entropy